# The Natural History of Stebbins Cold Canyon Reserve

## 2nd Edition

Edited by
Correigh M. Greene
and
Mikaela Huntzinger

Second editon edits by Shane Waddell

With contributions from:
Donald Shannon and Courtney Oqvist (Geology)
Laura Kindsvater and Kate Mawdsley (Plants)
Joel Ledford, Tammy Lim, and Barry Roth (Invertebrates)
Ron Cole and Ian Taylor (Vertebrates)
Peter Moyle (Hiking)

Photographs and illustrations by Jennifer French, Correigh Greene, Mikaela Huntzinger, Joseph Jacobs, Michele Johnson, Tammy Lim, Jake Mann, Sumudu Welaratna, and others

The Natural History of Stebbins Cold Canyon Reserve

Published by the University of California Natural Reserve System
Printed by DeHart's Printing Services Corporation

Front Cover: *Cold Creek* by Joseph Jacobs
Color Insets by Correigh Greene and Michele Johnson
Back Cover: *Cold Canyon Leaves and Stone* by Jake Mann



Second printing with corrections, 2004

ISBN 0-9714563-0-5

# Table of Contents

# Introduction

Stebbins Cold Canyon Reserve is one of the few natural reserves in the University of California Natural Reserve System that serves the dual purposes of academic research and public use. The 576 acres of the reserve are nestled in the Vaca Mountains, the eastern-most ridge of the North Coast Ranges. Cold Canyon and its neighbor to the southeast, Wild Horse Canyon, converge in the reserve and join the Putah Creek drainage just east of Lake Berryessa's Monticello Dam. The reserve's topography is primarily steeply sloping, with almost no flat land. The slopes of the canyon climb from 350 ft at the reserve entrance to 2437 ft at the highest point on the south rim. Furthermore, the reserve is almost entirely wild. A 1.5 mile hiking trail is all that connects the stone foundations of the old Vlahos homestead to Highway 128. As a result, Cold Canyon Reserve is both an excellent place for a day hike and a great location to study numerous aspects of California's natural history.

What exactly is natural history? The term "natural history" refers to at least two aspects of nature. First, natural history refers to the many and varied descriptions of the natural world. One type of description is a catalog of the plants and animals that inhabit Cold Canyon. This includes over 100 bird species, around 350 flowering plants, and many more insects, most of which remain undocumented. Another type of description is the shape of the landforms in Cold Canyon – the steeply sloping hills, the bedrock cliffs, and the boulder-strewn creek bed.

But species may come and go, as attested by the numerous non-native grasses that were tracked in by cattle and sheep in the late 1800s. And landforms change, sometimes in the sudden crash of a landslide, sometimes gradually through the action of water in the creek. Therefore, the concept of natural history also includes the processes that shape the natural world. For example, erosion – a very common geological process involving the movement of rocks and soil from the landscape – is integral to the formation of Cold Canyon and therefore a very important aspect of its natural history.

Both nature hikers and scientists can appreciate the complexity of the natural world, but their focus is often different. While a nature hiker might appreciate a general knowledge of the plants and animals he or she sees in Cold Canyon, a scientist might want to know the exact number of plant species occurring on a particular hillside. Because of these different levels of interest, this natural history field guide serves multiple functions. First, it provides general information about the reserve's rules, trails, and natural history for people interested in hiking in the reserve. The guide highlights common historical, physical, and biological features that you can easily observe in Cold Canyon if you are interested. Second, this field guide provides naturalists and researchers with detailed information about these topics, primarily in the form of species lists. More technical information can be found on the field guide's companion web site (http://nrs.ucdavis.edu/stebbins.html). As our knowledge of Cold Canyon's natural history continues to improve, the information detailed on the website will expand. The information contained in this guide, like that of all other natural history texts, is a snapshot of this knowledge.

# History of the Reserve

Stebbins Cold Canyon Reserve is one of 33 reserves in the University of California Natural Reserve System. Today's use of Cold Canyon by people as a research area and hiking spot spans only a short fraction of the canyon's entire human history. The dilapidated foundations of the Vlahos homestead in the center of the reserve remind us of the Canyon's relatively recent use as a ranch, but very few signs remain to tell us about the use of Cold Canyon by Native Americans dating back at least 4000 years.

## Native Americans

For thousands of years, Cold Canyon and the surrounding foothills and valleys were home to various Native American groups, most recently the Southern Patwin people. Until the first European settlers arrived near Cold Canyon, the Southern Patwin lived in semi-permanent villages and camps constructed of dome-shaped, subterranean thatch and mud houses. Common buildings in a settlement included a dancehouse, a sweatlodge, and an elevated granary to store preserved food. Individual homes were built for families, which consisted of a married couple and their unmarried children, as well as their married daughters and their husbands and children.

The Southern Patwin was a hunter gatherer society and used Cold Canyon to hunt rabbits, black bears, deer, tule elk, geese, ducks, and other game. They did not eat the feared but abundant grizzly bears of the area. Hunting practices were well developed and included setting low fires to encourage fresh plant growth, which, in turn, attracted more animals. In addition, they constructed long stone fences with a single opening to herd game so they could more effectively shoot at it. Although some of these now dilapidated and moss-covered stone walls can still be found in the hills nearby, none have yet been identified in the reserve. Arrowheads, another remnant of their hunting practices, have been found in the vicinity of Cold Canyon along Putah Creek. These arrowheads are important clues to the canyon's history and should not be removed.

The Southern Patwin did not farm but instead relied on the tremendous variety of edible plants found in the region, most of which can be seen from the trail as you hike through the canyon (see Table 1). The acorns from oak trees were the primary community staple. They ground these acorns into flour using mortars that they carved into boulders and using a bone or a rock as a pestle. At least one mortar exists in the bedrock along Cold Creek near the creek crossing, a permanent reminder of the native people that once dominated the area.

**A bedrock mortar along Cold Creek is the only remaining evidence that the Patwin once inhabited Cold Canyon.**

**Table 1.** Woody and non-woody plant species used by the Patwin for food, tools, craft supplies, medicine, and spiritual use.

| Species | Food | Tools |
|---|---|---|
| **WOODY PLANTS** | | |
| California buckeye (*Aesculus californicus*) | Seeds leached of poison, then used in soups. | Wood used for starting fires. |
| Bigleaf maple (*Acer macrophyllum*) | | Shredded bark made into aprons or skirts. |
| Foothill pine (*Pinus sabiniana*) | Pine nuts gathered and eaten. | |
| Interior live oak (*Quercus wizlizeni*) | Acorns were a staple. | Wood used to build shelters, paddles, and weapons. |
| Parry manzanita (*Arctostaphylos manzanita*) | Berries used for soup meal. Leaf stems brewed. | Wood used for making clubs. |
| Buck brush (*Ceanothus cuneatus*) | | |
| Western redbud (*Cercis occidentalis*) | Pods cooked and eaten. | |
| Mountain mahogany (*Cercocarpus betuloides*) | | Wood used for drills, arrow shafts, and throwing sticks. |
| Poison oak (*Toxicodendron diverilobum*) | Leaves used in lining cooking pits. | |
| Sandbar willow (*Salix exigua*) | | Twigs used in constructing sweathouses and fish traps. |
| Blue elderberry (*Sambucus mexicana*) | Elderberries eaten. | |
| California bay (*Umbellaria californica*) | | |
| California wild grape (*Vitis californica*) | Grapes eaten. Grape leaves used to line oven pits. | Vines used to lash granaries and fish traps. |
| **HERBACEOUS PLANTS** | | |
| Maidenhair fern (*Adiantum jordanii*) | | |
| Paintbrush (*Castilleja roseana*) | | |
| Soap plant (*Chlorogalum pomeridianum*) | Root eaten. | Extract used to stun fish in still waters. |
| California golden poppy (*Eschscholzia californica*) | | |
| Miner's lettuce (*Claytonia perfoliata*) | Leaves eaten. | |
| Wild oat (*Avena fatua*) | Grain gathered and pounded into flour. | |

| Craft | Medicine | Spiritual |
|---|---|---|
| | Pulverized seeds used as a poltice. | Branches protected against werebeasts. |
| Leaves used to line baskets. | | |
| | Pitch used as a bandage on sores and burns. | Pitch and tallow smeared on widows in mourning. |
| Soot from burned galls used for tattooing. | | Wood used to make drums. |
| | Steeped leaves helped cure poison oak and diarrhea. | Wood produced intense heat for ceremonial fires. |
| Switches used in basketry. | | |
| Bark and shoots used for pattern work in basketry. | | |
| | | Bark warded off thunder. |
| | Juice from pods used as a cure for warts. | Pods used for rattles in ceremonial dances. |
| Twigs used as foundations for baskets. | Bark brew used as a relief for poison oak rash. | |
| Stems made into rattles and flutes. | | Sticks tied in hair during religious ceremonies. |
| | Leaves used as cure for headaches and rheumatism. | |
| | | |
| | | |
| Stems used to weave black basketry patterns. | | |
| | | Leaves brewed as a love charm. |
| | Roots used in poultices. Extract used as soap. | |
| | Fresh root relieved toothaches. | |
| | | |

## HISTORY

Records kept by settlers indicate that as recently as 1838, the Patwin still lived along Putah Creek in large numbers. In 1848, with the end of the Mexican American War, California was ceded to the United States. California received statehood in 1850, and in 1852 the area that included Cold Canyon was legally defined as unappropriated unreserved public land. By 1877 the Southern Patwin were gone, forced out by Mexican and Spanish settlers who raised stock on their land-grant ranches.

### The Homestead

The Homestead Act of 1862 defined the region as stock raising land in 1916, and in 1938 John Vlahos was granted land in Cold Canyon for goat and cattle grazing. In order to raise collateral for his mortgage of $2500, he made 2000 pounds of goat cheese and built a cold storage building in a cool grove by the creek to store it. The cold storage building, whose stone foundation still stands and can be found at the terminus of the homestead trail, serves as the origin of the Cold Canyon name. The cold storage, the stone foundation of the Vlahos' house, a well, and some rusted remnants of an old vehicle are all that remain of the homestead.

**The remains of the Vlahos' cold storage.**

### Monticello Dam

Immediately upstream of Cold Creek's outlet to Putah Creek stands the massive wall of Monticello Dam. In the early 1900's, Berryessa Valley (the area now underneath Lake Berryessa) was a flat, fertile valley bisected by Putah Creek. The town of Monticello (population 250) stood in the center of the valley and was surrounded by 12,000 acres of farmland. Like the rest of Napa and Solano counties at the time, Berryessa Valley was farmed without the aid of irrigation. As farmers in many parts of the Central Valley constructed massive irrigation systems for their crops, dry-farmed produce quickly became uncompetitive in the market.

The farming interests in the region began championing dam construction to make local produce competitive. The construction of a dam at the narrow point of the Putah Creek Canyon, known as Devil's Gate, was first proposed in 1916, but several

Devil's Gate before the construction of Monticello Dam. Cold Canyon is near center on the left edge, and Berryessa Valley and Putah Creek can be seen beyond the "gate."

other dams were proposed on both Putah and Cache creeks in the ensuing decades. A proposal by the Solano Water Council to build a 304-foot dam at Devil's Gate became increasingly populara and the U.S. Bureau of Reclamation took over control of the project in 1945. The purposes for this dam were to (1) store water in Berryessa Reservoir for agricultural, urban, industrial, and military uses, (2) reduce flooding of lands along lower Putah Creek, and (3) provide recreational benefits. As construction of Monticello Dam became increasingly likely, the Winters Farm Bureau Center, the Yolo County Chamber of Commerce, the Winters Service Club, and the Yolo County Board of Supervisors each raised opposition to the project. Such opposition might today be phrased in terms of altered water regimes or disrupted plant and animal communities. However, in the mid-1940's, concerns about the dam centered around water prices and fair distribution of water to various groups. Local groups also opposed both the displacement of Monticello residents and the building of a dam in an earthquake zone. At the last minute, opposition by Solano County farmers and state authorities caused the height of the dam to be reduced from 304 to 270 feet, which correspondingly reduced the resevoir capacity from 2.2 to 1.6 million acre feet of water. The Secretary of the Interior approved the Solano Project, as the dam construction was known, in 1948. By 1957, the dam was complete, and by 1963, Lake Berryessa had filled completely.

Construction of the dam has probably affected use of Cold Canyon by animals. A dam and highway across Devil's Gate effectively barred stream-associated animal species from dispersing between Cold Canyon and regions upstream. No good records of vertebrate populations in the region exist before construction of the dam, but it is possible that construction of the dam, the reservoir, and Route 128 has isolated Cold Canyon from regions to the north and west.

Construction of Monticello Dam.

It is also possible that low flows created by limited water releases from the resevoir have eliminated some of the aquatic species that may have seasonally used the creek. Until recently, Putah Creek below Solano Dam (further downstream) was provided with just a small amount of water during the dry season. As a result, most of the creek dried up in the summer of 1989 (a drought year), killing thousands of fish and withering riparian vegetation. Similar events may have eliminated some species that swam into Cold Creek during the winter.

## Creation of the Reserve

John Vlahos sold part of his land to Paul Leiter in 1968. The Leiters sold this portion of the reserve to the University of California in 1979, and Petro and Virginia Vlahos sold the remainder to the University in 1984. Named in honor of Dr. G. Ledyard Stebbins, Professor in the Department of Genetics at U.C. Davis, the reserve was appropriated to preserve the land for teaching and research.

**G. Ledyard Stebbins in Cold Canyon.**

One goal of the Unversity of California Natural Reserve System is to preserve representative sites of each of California's many unique ecosystems. Stebbins Cold Canyon Reserve is an excellent example of Inner Coast Range communities because it retains many of its pre-European characteristics. With the obvious exception of large mammals such as grizzly bears and tule elk, the plant and animal communities of Cold Canyon seem largely the same as they have been since the first Native Americans arrived in this area. And although the trail and the remains of two structures dot the landscape, the canyon remains relatively free of development.

Still, recent human activity has left a strong mark on the landscape. The hills above the homestead trail are ribbed with the trails created by many years of livestock grazing, most recently the Vlahos goat herd. During this century, natural fires were suppressed in Cold Canyon and the surrounding foothills. Eventually, so much dead wood had accumulated that major fires, such as one that swept through the reserve in 1988, became inevitable. The 1988 fire was actually set by the California Department of Fire (CDF) in an attempt to complete a wildfire that first began south of the reserve, spread south with north winds, and then turned and moved north with southwest winds. CDF decided to drop fire sticks in various locations to finish the burn between Highway 128, Pleasant's Valley, and Highway 80. CDF did a careful and meticulous job; only a few oaks burned and the canyon was unburdened of several tons per acre of fuels. Some of the patches in the hillside vegetation were created by "hot spots" in that fire. A more subtle but nonetheless significant effect of humans is the elimination of the grizzly bear as a major predator and earth mover in the area. In recent years, recreational users have created many unauthorized trails through the reserve and increased erosion of the official trails. The number of visitors to the canyon varies consistently with weather, but the trails are open all year round, and become well-worn after the winter rains. As you walk up the trail, keep these changes in mind and think of the effects of your own presence on plants and wildlife.

# The Physical Environment

While humans can modify the landscape within a few generations, the physical forces of nature have just as large an impact on the shape of the land, but occur on much slower time scales. Geologic processes may mold the land over millions of years. Annual variations in rainfall and temperature influence which plants and animals become established or go extinct in the area on a time scale calculated in thousands of years. Therefore, understanding Cold Canyon's physical environment helps explain the diversity of life we see in in the reserve today.

## Coast Range Climate

Northern California's climate is characterized by cool, wet winters followed by harsh, dry summers. Monthly changes in temperature and rainfall are therefore extreme and highly seasonal. This variation is caused by seasonal changes in weather patterns in the Pacific Ocean. During much of the year, a zone of high pressure running up the coast of California prevents the low-pressure storm systems that inundate the northwest coast from reaching much of Northern California. During the winter, this high pressure system breaks down, allowing the storm systems to move farther south. As a result, Cold Canyon receives precipitation primarily from November to March, much of it falling in the winter months. Accompanying the rain are cooler temperatures. During the winter, the thermometer often dips below freezing, especially during the night, and by day the temperature averages 50° F (10° C).

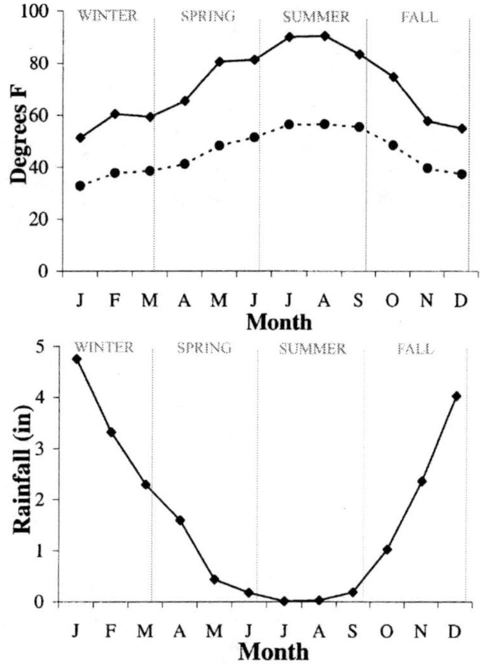


**Minimum and maximum mean daily temperature and monthly rainfall over 12 years, taken at Markley Cove (1.5 miles west of Cold Canyon).**

The situation is reversed in the summer, when daily temperatures can soar above 100° F (38° C) and less than 1% of the year's rain falls. In some respects, Cold Canyon is even harsher in the summer than other areas of the coast range. Nestled within the easternmost ridges of the North Coast Range, Cold Canyon receives somewhat less rain than hills to the west. These hills also prevent the drought-alleviating coastal fog from reaching Cold Canyon most of the summer. Furthermore, the canyon traps heat and the southern and western ridges shield the canyon from the cool delta breezes that blow in from the coast in the afternoons.

## The Geology of Cold Canyon

Like the entire surface of the Earth, the landscape of Cold Canyon is constantly changing, albeit at a very slow rate. Geologic processes are continually reshaping the rocks and sediments that make up the landforms on the surface of the planet. Less than one million years ago, Cold Canyon was merely a shallow trough, not the deep canyon present today. Millions of years before that, the rock that would later form Cold Canyon was located far beneath the surface of the sea. The present landforms of Cold Canyon are the products of three slow but very active geologic processes: deposition of sediments, continental uplift, and erosion. Over many millions of years these geologic processes have created, uplifted, and eroded vast amounts of rock to create Cold Canyon. Evidence of all three processes can be seen today at the mouth of the canyon. Sedimentary layers bent skyward by uplift surround Monticello Dam, while nearby, Cold Creek erodes material from the canyon walls and deposits it on the banks of Putah Creek.

**Near-vertical layers of sandstone in the ridge north of Monticello Dam.**

**Sediment deposition on the banks of Putah Creek east of Cold Canyon.**

### *Sediment Deposition and Bedrock Formation*

Millions of years ago, off the western coast of the ancient continent, the land that would later become Cold Canyon was being deposited as river delta sediment. As rivers flow into the sea, they lose velocity, and thus lose their ability to carry sediment. Larger, heavier particles carried by the river, such as sand, are deposited close to shore, while finer-grained particles, such as silt and clay, are carried farther offshore by the diminishing current. Some sediment is carried all the way to the continental shelf, the edge of the continent where it falls away steeply into the sea. Here, sediments accumulate on the steep slope, and over time become unstable. These sediments can then slide under their own weight, similar to a landslide or avalanche. These "undersea landslides" are called turbidites and can occur quite regularly. As the muddy turbidite slides down the continental slope, the larger, heavier particles, such as sand, are the first to settle onto the slope, followed by lighter, finer particles, such as silt and clay. In the end, the muddy sediments have been sifted into layers, with silts and clays generally overlaying sands. This layering can occur at two scales: beds of silt over 200 feet in depth may overlay similar-sized sandy deposits, and within a single bed (several inches thick), sediment grain size changes from sandy material at the bottom to silty material at the top. Geologists use these small-scale changes in particle size to determine the original position of different layers relative to each other.

Over several million years, as new depositional layers are built upon old layers, the weight of the upper layers creates great pressure on the lower layers, causing a trans-

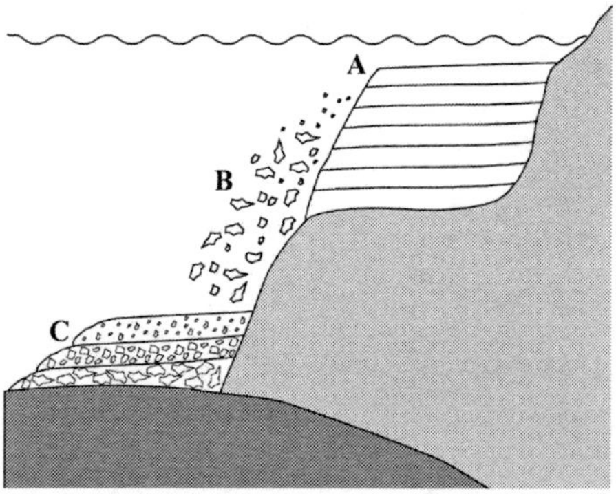

A. Sediment deposition in a delta. B. An undersea landslide on the steep edge of the delta. C. Sediment is deposited into layers of different grain size, with finer layers on top.

formation from muddy sediment to solid rock. The coarse sediments form sandstones, while the finer particles form mudstones or shales. These types of rocks can be seen today in Cold Canyon. The canyon walls, including the rock outcrops on both ridges, are predominantly sandstone. The bedrock underlying the canyon bottom is shale. These bedrock layers have been given names and approximate ages. Pleasants Ridge, the eastern ridge of Cold Canyon, is part of the Sites Formation and was deposited 88 million years ago. Blue Ridge, the western ridge, is part of the 91 million-year-old Venado Formation, and the shale canyon bottom is part of the Yolo Formation, which was deposited 90 million years ago.

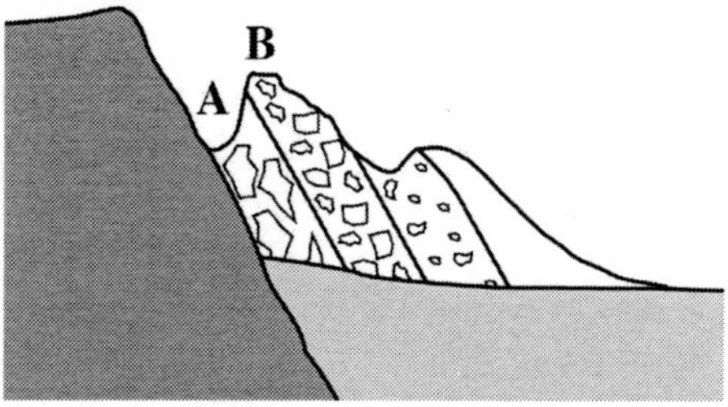

Continental uplift has changed the orientation of the sediment layers, and erosion has cut valleys into less resistant layers (A) and leaving behind more resistant layers (B) as ridges.

## *Continental Uplift*

As deposition was occurring millions of years ago, the entire continent was slowly rising. The entire surface of the planet is covered by a number of massive plates, rafts of solid rock afloat on a layer of molten rock. These plates are in constant motion, jostling and grinding against each other. For example, the San Andreas fault, which runs through much of California and causes many earthquakes, is the boundary between two plates which are sliding past each other. Another type of plate boundary, a convergent boundary, occurs when two plates are moving toward one another. When the plates collide, the denser oceanic plate typically dives beneath the other and into the interior of the Earth. The lighter continental plate rides on top and is slowly uplifted as a result.

California's Coast Range was formed by the actions of a convergent plate boundary. A collision of the oceanic plate with the continental plate forced the coastal sediments up above sea level. Further pressures from this collision bowed these horizontal sedimentary layers into a near vertical orientation. This vertical layering is visible around Monticello Dam, near the mouth of Cold Canyon. After uplift, the motion along the plate boundary changed from convergence to lateral sliding, as attested by the motion of the San Andreas Fault today.

A mineral spring on the west face of Cold Canyon.

The Cold Canyon region may still be tectonically active. To the north of Putah Creek, a number of mineral springs lie scattered on a line extending southward towards Cold Canyon. The presence of certain minerals and the linear pattern of springs on the landscape indicate the presence of a fault in this area. A similar spring is located within Cold Canyon. Although the evidence is not conclusive, the presence of this spring suggests that an active fault runs through the canyon, and within a quarter mile of Monticello Dam.

## *Weathering and Erosion*

Cold Canyon formed as a result of weathering and erosion of the different bedrock layers by wind and water. Weathering is the breakdown of rock into small particles and can occur through chemical or physical mechanisms. Erosion is the transport of these particles by running water or landslides. Running water will typically seek the path of least resistance, cutting through softer, finer-grained rock layers, while avoiding more resistant layers. This is evidenced by the fact that the floor of Cold Canyon lies in the midst of the Yolo Formation, which is a fine-grained mudstone layer, while the canyon walls consist of more resistant Sites and Venado Formation sandstone layers.

The erosion that created Cold Canyon continues today both on a small scale, through the chemical breakdown of rock by air and rain, and on a larger scale by physical mechanisms such as landslides. Landslides are common on the steep walls of Cold Canyon. In the 1980's, two different slides covered the main trail. In 1995, a large landslide came down the eastern wall, toppling trees and piling soil and rocks along the streambank. The evidence of this slide can easily be seen along the trail, near the mouth of the canyon where the trail comes closest to the creek.

However, these landslides were tiny compared to a massive slide that must have occurred in Cold Canyon many thousands of years ago. This slide formed the entire western slope of the canyon, and exposed the rock outcrops topping Blue Ridge. To best appreciate the size of this slide, hike up the western trail. Near the top of the trail, you will be able to look at rock outcrops on the ridgetop extending to your left and right. The large cliff on the left is the bedrock of the Venado Formation that was exposed by the massive landslide. In fact, the slope you are standing on, which runs south underneath the cliff and along the rest of the canyon, is the top of the huge mound of earth that slumped several hundred feet in that landslide!

**Remains of the 1995 landslide near the reserve entrance.**

**This rock outcrop is actually the scarp of a huge landslide!**

More gradual erosive processes are also currently taking place. The creek actively erodes its channel, especially during winter floods, and Cold Canyon gets slightly deeper every year. The rounded boulders in the creek bed, smoothed by the scouring action of sediments transported by the stream, attest to this on-going erosion.

## *A Geologic Puzzle in Cold Canyon*

Geologists dedicate their work to interpreting the processes that formed the landscape features that we see today. This task can be particularly daunting, as most geologic formations are completely covered by other rock formations, soil, and plants. The landscape of Cold Canyon can be a particular challenge – the many landslides that have occurred in the canyon have hidden much of the underlying bedrock that might yield clues to ancient geologic events. Fortunately, geologists have found enough outcrops in Cold Canyon and other neighboring areas to provide the general geologic history described above.

Sometimes, however, the incomplete information available creates a geologic puzzle. One such puzzle exists in Cold Canyon. This puzzle concerns the orientation of a particular bedrock outcrop. Recall that the bedrock layers were created long ago by the filtering action of water. Within a particular bed of sediment, coarse material settled first, followed by finer and finer layers. For example, the sandstone outcrop near the trail's creek crossing reveals that rock layers composed of coarse-grained particles lie underneath finer-grained layers. At this outcrop at least, the rock layers still lie in relatively the same way they were laid (and "fine-upwards").

**Fining-upward beds near the creek crossing.**

It is thus rather surprising to discover a very different pattern in an outcrop just several hundred meters north of the outcrop at the creek crossing. This second outcrop occurs on the west side of the creek near an old landslide scar. Checking the gradation in particle size in the layers of this outcrop reveals that the coarser layers occur above the finer beds. The entire outcrop is upside-down! How could these layers of rock become overturned? This, in essence, is the geologic puzzle of Cold Canyon.

**Overturned beds.**

An answer to this puzzle is not easy to obtain, because the clues that would help solve the mystery are buried deep beneath the western canyon wall. However, there are at least two plausible explanations for the existence of these overturned beds. A relatively simple explanation supposes that some of the bedrock layers, moved in the huge landslide that shaped the western canyon wall, were overturned as they came crashing down the hill side. A second, more elaborate explanation alludes to the existence of a large fault in the canyon. In this scenario, the Venado Formation was split into two sections by a fault, and then one section was thrust above the other by continental uplift. This thrust may have caused certain layers caught near the edge of the fault to become overturned.

One clue supporting the second explanation is the existence of the mineral spring, less than 660 feet (200 m) from the location of the overturned beds. Because mineral springs tend to follow fault lines, and are less likely to occur within the jumble of landslide debris, the existence of the spring suggests that a fault line lies very close to the overturned beds.

Gathering clues like the presence of the spring is the key to understanding the geologic processes that shaped Cold Canyon. Hence, the present landscape can be used to explain the actions of the past. At the same time, an improved understanding of the link between past events and present geologic features can help geologists predict what might happen in Cold Canyon in the future. For example, if additional observations help connect the mineral spring in Cold Canyon with those in areas to the north and lead geologists to determine that an active fault runs through Cold Canyon, we might conclude that the bedrock along this fault is particularly unstable. We would therefore predict that any earthquakes in the region could be centered along this fault line.

# Vegetation of Cold Canyon

Stebbins Cold Canyon Reserve provides an excellent opportunity to experience vegetation characteristic of the Inner Coast Ranges of North-Central California. Here, the variable California climate produces dramatic changes in vegetation, both in time and space. Over the course of the year, rain and temperature cycles force many plants to grow and reproduce during specific seasons. In addition, the relatively small area of Cold Canyon harbors a variety of plant communities, including grassland, savanna, chaparral, live oak woodland, and riparian (riverside) woodland. These communities offer a wealth of plant life. This chapter discusses how climate influences which plant species occur in Cold Canyon, the communities in which these plants live, and what common species you can easily observe in Cold Canyon. Since many of these species produce beautiful flowers during certain months, this chapter concludes with a blooming "calendar" of common plants (p. 36). A complete list of plants known to occur in Cold Canyon can be found in the appendices (p. 95).

## Coping with the Climate

Much of California enjoys the comfortable weather associated with Mediterranean climates, which are characterized by cool wet winters and hot dry summers. Humans love these climates and the plants associated with them. However, these extreme weather conditions pose quite a challenge for the plants that live in California, particularly in Northern California, where the benefits of the Mediterranean climate are fewer and the hardships are greater. The cool winter temperatures retard much plant growth in Cold Canyon, and many plants do not grow at all in this season. By December, many trees have shed their leaves, and most other perennial plants have shut down their growth. A number of plants have extensive bulb or root systems that store nutrients during the winter so that the plant can resprout in the spring.

However, the greatest challenges to the plants that inhabit Cold Canyon are the summer temperatures and lack of rain, which turn the region into a temporary desert. Many native grasses avoid the harshness of summer by growing and producing seeds during the wet months of winter. They use the available water to grow and reproduce through the winter, and sit, brown and dormant, through the summer. Many annual plants have evolved a similar strategy, growing only during the mild and wet months of late winter and early spring. Therefore, spring is the ideal season to see many of Cold Canyon's beautiful wildflowers.

To survive through the summer drought, some plants live in cool "microclimates." The majority of these species avoid drying out by growing in the moist and shady canyon bottom along Cold and Wild Horse creeks. These areas are much cooler and moister than the average conditions of the reserve. Experience this microclimate by hiking to the location of the cold storage foundation at the end of the main trail. Even on a hot summer day, this spot is remarkably cool. Mild microclimates exist in wet areas, in shaded areas, and on north-facing slopes, while the harshest microclimates exist in rocky, bare, sunny areas on south-facing slopes.

Perennial species that do not rely on riparian microclimates often have adaptations that enable them to retain or gather moisture. Some plants found in Cold Canyon are sclerophyllous

**Pleasants Ridge, to the east of Cold Canyon. The northern slopes are chaparral, while the southern slopes are grasslands.**

("having hard leaves"), and have small leathery leaves covered with a thick waxy coating that minimizes water loss. Other plants grow extremely deep root systems to find water. One such example is California's state grass, purple needlegrass (*Nassella pulchra*), which sometimes grows roots extending 13 feet (4 m) into the soil.

**The sclerophyllous leaves of chamise.**

## Plant Communities

Because both mild and harsh microclimates exist in the reserve and because different plant species are more or less tolerant of these microclimates, certain species tend to live in similar places. These groups of co-occurring species are known as plant communities. Plant communities are not easily defined, as many plants occur in different communities. Nonetheless, botanists have identified five general plant communities in Cold Canyon. These are termed grassland, savanna, chaparral, live oak woodland, and riparian woodland.

**A patch of grassland surrounded by chaparral.**

*Grassland*. These areas are characterized by very little shrub or tree cover, and instead are blanketed by annual or perennial grasses and other annual plants. In the spring, grasslands are often covered by Cold Canyon's many wildflowers, but most of these plants die before the hot, dry weather. Grasslands are most frequently found on south-facing slopes, which are not readily colonized by shrubs and trees because of the harsh summer microclimate. Grasslands also occur in areas that have recently experienced a natural disturbance such as a fire or landslide.

In these areas, the grasslands are more temporary and will eventually be replaced by shrubs and trees.

Grassland communities of Cold Canyon and throughout the West have changed dramatically in the last few centuries. When European settlers came to California, they brought many new grasses and forbs (non-woody plants, such as mustards, thistles, and clovers) with them, and these mostly annual exotic species now dominate the herbaceous vegetation of Stebbins Cold Canyon Reserve, particularly its grasslands.

*Savanna.* This association is characterized by scattered trees with an understory of grasses and forbs. Savannas exist in relatively dry microclimates; if more water were available, more trees would be able to grow, and the area would become a woodland. Therefore, the dominant trees of savannas can tolerate drought. Trees of Cold Canyon's savanna include the blue oak (*Quercus douglasii*) and the foothill pine (*Pinus sabiniana*).

**Savanna on the north end of Pleasants Ridge.**

*Chaparral.* Chaparral, the most common community type in Cold Canyon, is a dense mixture of shrubs that range in height from 3 to 9 feet (0.9-2.7 m). In some places in Cold Canyon, the chaparral community is entirely dominated by chamise (*Adenostoma fasciculatum*), which produces a substance from its roots that appears to prevent many other plants from growing in the same soil. Other common chaparral shrubs of Cold Canyon include parry manzanita (*Arctostaphylos manzanita*), scrub oak (*Quercus berberidifolia*), toyon or Christmas berry (*Heteromeles arbutifolia*), and buck brush (*Ceanothus cuneatus*).

**Dense chaparral in the foreground.**

*Live oak woodland.* In cooler hillside areas where more water is available, live oak woodland predominates. Live oak woodland is characterized by a relatively dense growth of trees and shrubs, which provides a fair amount of shade. This plant community is dominated by interior live oak (*Quercus wislizeni*), the evergreen oak of Cold Canyon, but also includes foothill pine, blue oak, manzanita, buck brush, California coffeeberry (*Rhamnus californica*), and California buckeye (*Aesculus californica*).

**Live oak woodland along the trail up to Blue Ridge.**

## VEGETATION

*Riparian woodland.* The word "riparian" comes from the Latin word meaning "river bank." Thus, riparian woodlands are the forests associated with streams and springs. Cold Canyon's band of riparian species is limited and narrow, due primarily to soil and topography. This community type includes trees associated with other communities, such as California buckeye and interior live oak. A number of less drought-tolerant species, such as Fremont cottonwood (*Populus fremontii*) and various willows (*Salix* spp.), grow right at water's edge, sometimes becoming surrounded by the creek as it shifts course due to landslides or torrential rainstorms. California bay (*Umbellularia californica*), with its aromatic leaves, and Western redbud (*Cercis occidentalis*), with its bright magenta blossoms that surprise March visitors, are both scattered along the creek. Groves of the bigleaf maple (*Acer macrophyllum*) provide shade in the upper regions of Cold Creek.

**Riparian woodland.**

## Common Woody Plants

**Bigleaf maple** (*Acer macrophyllum*). Maples are easy to identify in spring and summer when they have leaves. As its name implies, the bigleaf maple's leaves are quite broad, and they are incised into several pointed lobes. There are many old maple trees growing by the cold storage of the old homestead. The paired fruits of the maple are winged for distribution via wind and have entertained countless children with their helicopter-like flight. Native Americans lined their baskets with the tree's large and graceful leaves. They also shredded maple bark to make aprons or skirts, the standard attire for women, as well as to make diapers and to wrap the dead before burial.

**California bay** (*Umbellularia californica*). The California bay has bunches of small yellow flowers in winter and spring and elongated leaves that smell strongly aromatic when crushed. Also known as peppernut, the fruits of this plant are edible. The Patwin may have dried them, roasted or parched them in a fire to make them less bitter, and then cracked and ate them. Early Californians discovered many medicinal uses for this plant. They used it to treat headaches, stomach aches, and rheumatism, and laid it in their dwellings to repel fleas. Athough the leaves of this plant are not those sold in supermarkets as "bay leaves," they make a marvelous substitute.

**California buckeye** (*Aesculus californica*). California buckeyes range in size from 6 to 30 feet (1.8-9.1 m) in height. In winter the white branches of the California buckeye point dramatically skyward, barren of leaves and contrasting starkly to evergreen shrubs. Early in the spring, it unfurls bright green compound leaves composed of six to eight leaflets. Later in the spring, it produces large, showy, spike-like white or pinkish stems of flowers that cover its canopy, giving it the appearance of a great candelabra. The buckeye loses its leaves in August or early September, probably as a response to drought. Once its leaves fall, the pear-shaped buckeye fruits hang prominently from the ends of the tree's branches. These fruits open up in response to the fall rains and drop their huge seeds. The buckeye seed is the largest of any tree in western North America. The seed is a beautiful mahogany brown, and it harbors a toxin so powerful that Native Americans used a ground extract to paralyze fish.

Bigleaf maple

California bay

California buckeye

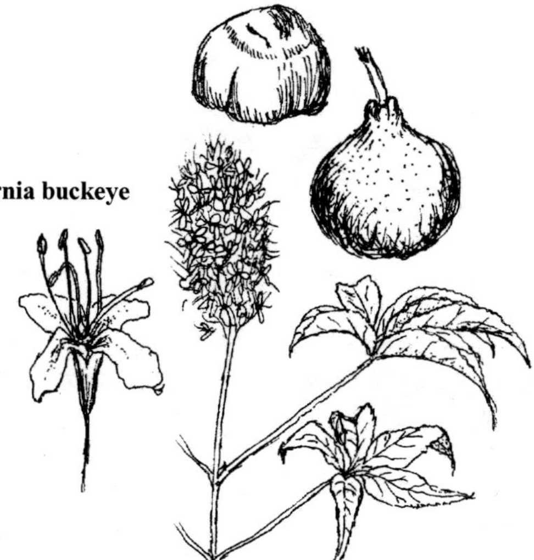

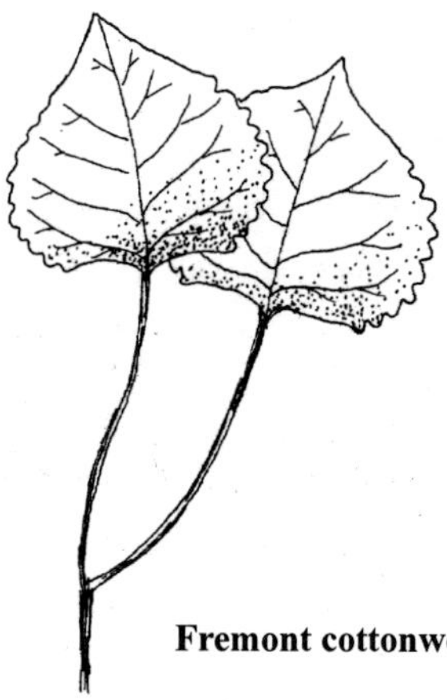

Fremont cottonwood

Foothill pine

## VEGETATION

**Foothill pine** (*Pinus sabiniana*). Foothill pine, also called ghost or gray pine, is the only conifer in Cold Canyon. The foothill pine is unusual not only for its ability to live in hot, dry areas, but also for its branched appearance. The large seeds of the foothill pine were an important source of food for local tribes.

**Fremont cottonwood** (*Populus fremontii*). Cottonwoods grow to be large trees with coarsely grooved, grey bark. Their leaves are wide and somewhat heart-shaped, with teeth along the margin. The trees require lots of water and grow right at water's edge or sometimes even in the stream. The name "cottonwood" is derived from their downy seeds, which form snowdrift piles of fluff in late summer.

**Oaks** dominate Cold Canyon, and are very important for its ecology. They provide shade for numerous other plants. Their wrinkled bark houses many species of invertebrates, and because oaks have a tendency to become hollow without dying, they also provide homes to numerous cavity-seeking animals. In addition, their acorns – available during winter when other food is scarce – provide an important food source for animals as diverse as quail, squirrels, chipmunks, deer, and black bears.

Acorns were also a staple food for the Native Americans of this region. A skilled harvester could gather enough acorns in just five days to last a year. Often the Native Americans ground the acorns into flour, washed the resulting meal to leach it of bitter tannins, shaped it into cakes, and baked it. Acorns contain a significant amount of calcium, phosphorus, magnesium, potassium, sulfur, fat, and protein, nutrients that are often scarce and valuable.

**Blue oak** (*Quercus douglasii*). Although bare of foliage in winter, blue oak can be identified in spring and summer by its highly variable, pastel-green leaves. Some leaves are deeply lobed, others are not lobed at all; some have gently curved edges, others have almost prickly edges. Of the three tree oak species living in the reserve, blue oak is probably the species whose acorns were most frequently gathered by Native Americans.

**Interior live oak** (*Quercus wislizeni*). Unlike blue oak, interior live oak is evergreen, with shiny, prickly leaves. It is sometimes difficult to tell interior live oak from scrub oak, although scrub oak can usually be identified by its shrubby appearance and more wrinkled bark. Native Americans of this region preferred the acorns of blue oak or valley oak, but they also gathered the acorns of live oak in times of emergency.

**Scrub oak** (*Quercus berberidifolia*). Scrub oak is a bushy shrub with evergreen leaves that have abruptly pointed teeth. It is difficult to distinguish this oak from the shrub form of interior live oak because both oaks are evergreen, but the bark of the interior live oak is much smoother.

**Valley oak** (*Quercus lobata*). The acorns of valley oak, which grows nearby along Putah Creek, may have been even more prized by Native Americans than those from blue oak. Valley oak leaves are greener, larger and more deeply lobed than those of blue oak. There is one valley oak growing just outside of the entrance to the reserve, and several others grow further up the creek near a permanent spring.

Blue oak

Interior live oak

Scrub oak

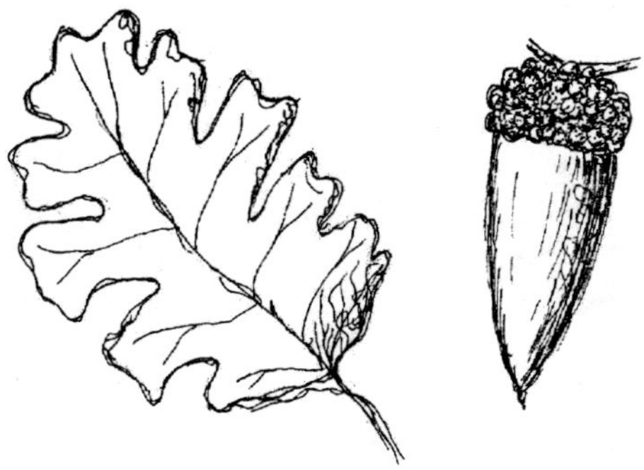

Valley oak

Spice bush

Blue elderberry

## VEGETATION

**Spice bush** (*Calycanthus occidentalis*). Crush the leaves of spicebush for its distinguishingly strong, sweet smell. The unusual flowers of this plant, made up of numerous, spirally arranged petals and sepals, are bright pinkish red and measure approximately 2 inches (5 cm) in diameter. The leaves can be quite long, up to 6 inches (15 cm) in length.

**Blue elderberry** (*Sambucus mexicana*). Blue elderberry is a tall shrub that grows in moist areas. Its leaves are pinnately compound and opposite one another. In April, elderberries display clusters of small cream-colored flowers, which develop into purplish-black, round fruits. The fruit is toxic unless cooked. This shrub is the larval host plant for the reserve's only known federally listed endangered species, the valley elderberry longhorn beetle (*Desmocerus californicus dimorphus*).

**Buck brush** (*Ceanothus cuneatus*). This is one of two "California lilacs" growing in the reserve, which produce delicate plumes of blue flowers in the spring. Buck brush can be identified by the presence of three distinctive veins branching from the base of each leaf.

**Chamise** (*Adenostoma fasciculatum*). Probably the most common chaparral plant in the state of California, chamise is a dense shrub whose branches are lined with short (less than 1 cm), needle-like leaves. The low surface area and hard coating on these leaves help prevent water loss during the harsh summers. In late spring, chamise shows its small five-petaled white flowers. Chamise is highly adapted to fire and will resprout quickly from the roots when burned. The shrub has an interesting strategy called allelopathy, of excreting toxins into the ground, apparently to reduce competition for water and nutrients by other plants.

**Coyote brush** (*Baccharis pilularis*). Coyote brush is a common chaparral plant found in the reserve. It is readily distinguished by its leaves, which are smooth and thin and have irregular edges. Coyote brush, and its close relative mule fat, are particularly interesting plants because they are dioecious – different plants are either male or female, but never both. In constrast, most other plants are monoecious, which means that all plants bear flowers that produce both seeds and pollen.

**Mule-fat** (*Baccharis salicifolia*). This plant looks very similar to a willow. Its easy to mistake it for one, unless it happens to be blooming, and then its wispy, fluffy seed heads reveal its true identity as a relative of the common dandelion. Mule-fat grows along stream banks as well as in the creek.

**Parry manzanita** (*Arctostaphylos manzanita*). Manzanita is a shrub with smooth, reddish bark that peels off in strips. The leaves of manzanita are thick, light green, paddle shaped, and between 0.5 and 1.5 inches (1-4 cm) in length. Like those of toyon, manzanita's berries are red in color and a good source of vitamin C. Native Americans made a drink by removing the skins and seeds of the berries, then grinding the pulp into a powder, mixing it with water, and letting it stand for several hours before drinking it. They also dried the berries for winter use.

**Poison oak** (*Toxicodendron diversilobum*). Poison oak is ubiquitous in Cold Canyon at high densities. At any time of the year, this plant can cause severe skin reactions for some people. Hence, it is wise to know what this plant looks like and to avoid contacting it. Hiking off-trail

Buckbrush

Coyote brush

Chamise

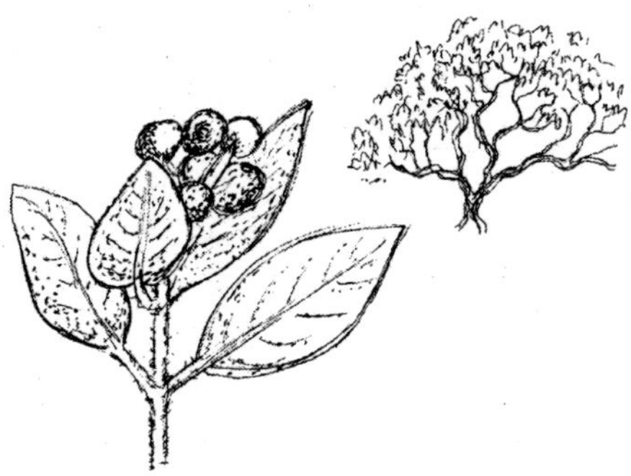

Parry manzanita

Mule fat

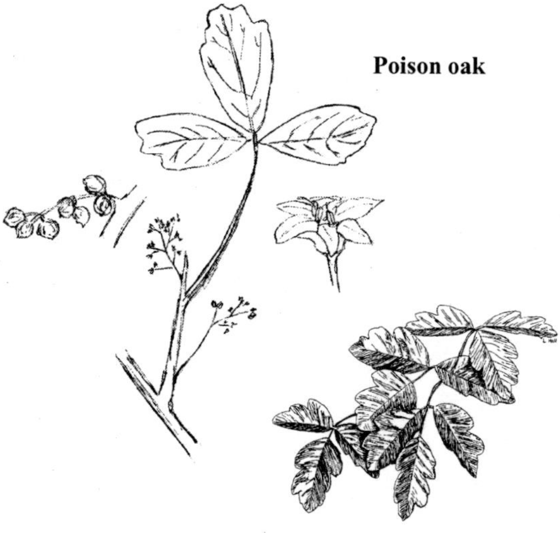

Poison oak

is a sure way to get covered in poison oak, and even on the trail, encroaching stems can touch you if you are not on the lookout. If your skin does come in contact with the plant, thoroughly rubbing the skin with an alcohol swab may remove much of the noxious oil before it takes effect. Regardless, wash the area with a strong soap as soon as possible. Keep in mind that the oil can get on clothes, shoes, and other surfaces, and be transferred to skin for a long time thereafter.

Learn to recognize this plant while hiking. Poison oak usually has the appearance of a small shrub or vine, with sets of three oily, green leaflets of varied shape and color. In the winter, poison oak's straight twigs can be identified by the "pointing-finger" shape of its buds. These buds sprout shiny red leaves in the early spring. At the same time, the plant produces catkins of white flowers. The red leaflets turn green after several weeks, and may look very similar to the lobed leaves of oaks, or may also have entirely smooth edges. However, these leaflets always grow in groups of three. In late summer and fall, the leaflets turn bright red before falling off. After it loses its foliage, poison oak can often be recognized by its clusters of green-brown or white berries that are a very important food source for wintering birds and other animals.

**Skunkbrush** (*Rhus trilobata*). This plant grows all along the homestead trail, scaring hikers with its close resemblance to poison oak. Skunkbrush, however, does not cause skin irritation. One method to tell the two plants apart is that the flowers of skunkbrush are located in a bunch at the end of each branch, whereas the flowers of poison oak come out from each side of the branch and not at the end.

**Toyon** or **Christmas berry** (*Heteromeles arbutifolia*). This shrub draws attention with its bright red berries and distinctive elongated, toothed leaves. The berries are bitter-tasting; Native Americans cooked them before eating. They also made a tea from the bark and leaves as a cure for stomach aches and other aches and pains.

**Western redbud** (*Cercis occidentalis*). Redbud thrives in a multitude of habitats in Cold Canyon – at streamsides, on dry slopes, and in chaparral. In early spring, these shrubby trees produce beautiful showy pink flowers. Thereafter, they produce heart-shaped leaves. By summer, the redbud's fruits, which are long dark brown pods, have fully developed and hang from the tree in clusters. The pods are reputedly edible. Native Americans also used the bark and shoots for dark pattern work in basketry.

**Willow** (*Salix* spp.). Different species of willow are very difficult to tell from one another, and there are at least four different species in Cold Canyon. They only grow close to or in the stream. Willows are famous in the ethnobotanical world for manufacturing the chemical compound from which aspirin was derived. Its thin yet strong branches had a variety of uses to Native Americans and were particularly important for basketry.

**Yerba santa** (*Eriodictyon californicum*). This shrubby plant with oily, thin, dark green leaves produces a cluster of white flowers that is irresistible to many animals, especially butterflies and hummingbirds. Yerba santa literally means "holy herb," and it was used by Native Americans for a variety of medicinal purposes, such as a treatment for stomach aches and rashes. Recently, yerba santa has been listed in pharmacological texts as a treatment for bronchitis.

Toyon

Skunkbrush

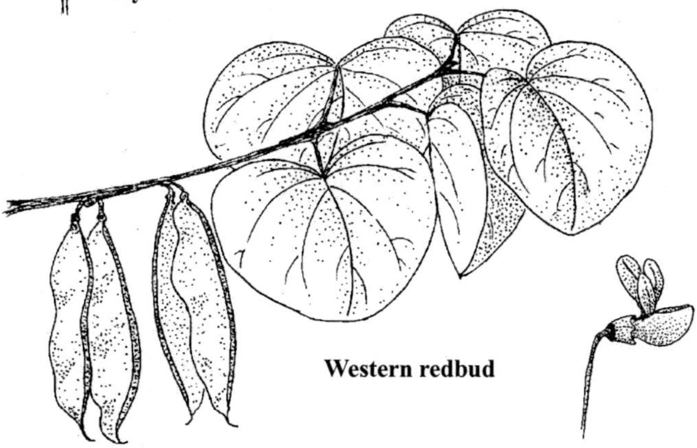

Western redbud

Willow

Yerba santa

## VEGETATION

# Common Herbaceous Plants

### *Ferns*

**Maidenhair fern** (*Adiantum jordanii*). Ferns abound along the banks of Cold Creek. Maidenhair fern has small, delicately shaped, rounded leaves with crinkly edges. Two centuries ago, the Patwin of Cold Canyon probably collected the black stems, weaving them into intricate basketry patterns.

**Giant chain fern** (*Woodwardia fimbriata*). The beautiful leaves of the giant chain fern grow to lengths of 3 to 10 feet (1-3 m). Native Americans also collected this fern, extracting the two white strands in the stem, dying them red with the bark of red alder, and then using them for red pattern work in basketry.

### *Grasses*

**Annual grasses**. European settlers brought many species of plants with them, either purposefully or inadvertently. European annual grasses such as wild oat (*Avena* spp.), foxtail (*Hordeum* spp.), and ryegrass (*Lolium* spp.) have flourished spectacularly and spread throughout the West. Spaniards probably planted broad-leaved wild oat (*Avena fatua*), but numerous California tribes, including the Pomo, Cahuilla, and Miwok, gathered it for food, grinding the parched seeds into meal. Waterbirds and songbirds also feed on the seeds of wild oat. Unfortunately, other introduced annual grasses, such as many of the bromes (*Bromus* spp.), cover large tracts of land in California but do little to provide food or shelter for native rodents, birds, insects, or large mammals.

**Perennial bunchgrasses**. California's grasslands were at one time dominated by mostly perennial bunchgrasses, such as woodland wild rye (*Elymus glaucus*), California melic grass (*Melica californica*), and foothill and purple needlegrass (*Nassella* spp.). All three of these grasses still grow in parts of Cold Canyon, but annual grasses from Europe outcompete them for space and soil resources, so they are no longer dominant. There are some beautiful examples of these perennials growing on a hillside just after the first major creek crossing.

### *Flowering Plants*

**Bush monkeyflower** (*Mimulus auriantiacus*). Monkeyflowers derive their name from the observation that the fused petals of the flower appear to form a monkey face. The bush monkeyflower can be found growing on rocky hillsides and chaparral edges. In spring and summer, look for buff yellow to salmon-orange flowers blossoming on small shrubs composed of smooth, sticky leaves. Native Americans made a medicinal tea from the leaves of this and other species of monkeyflower.

**California poppy** (*Eschscholzia californica*). Sprinkled about the rolling, grassy hillsides is a wide variety of spring wildflowers, including the California golden poppy, our state wildflower. Native Americans placed fresh poppy root in cavities to relieve toothache. The roots of California poppy are a startlingly bright shade of orange and look somewhat like miniature carrots.

**California pipevine** (*Aristolochia californica*). This vine can often be found curling up around poison oak and other shrubs. Pipevine is often the first plant in Cold Canyon to flower, opening

Maidenhair fern

California melic grass

Giant chain fern

Purple needlegrass

Wild oat

California poppy

Bush monkeyflower

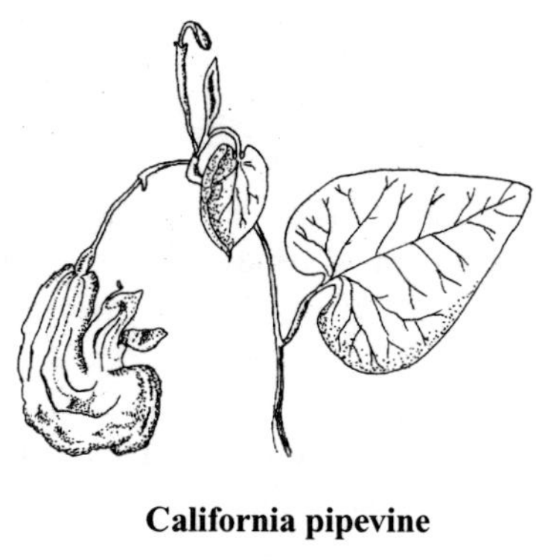

California pipevine

its oddly shaped, tube-like flowers as early as February. The sepals of this flower are colored green or light brown with a lining of pink or red thickened tissue and purple veins. The leaves are heart-shaped. Fungus gnats pollinate this plant, drawn by a metallic odor.

**Chinese houses** (*Collinsia heterophylla*). In early spring, Chinese houses can be found growing in shady areas of the reserve. Their pink or lavender flowers bloom in tiers of whorls around the stem, an arrangement that can look like the rooflines of Chinese pagodas. Native Americans used a poultice of leaves from *Collinsia* to treat insect and snake bites.

**Delphinium** (*Delphinium* spp.). There are at least three species of delphinium, also known as larkspur, living in the reserve, Most delphinium species tend to have beautiful deep bluish purple flowers, which are recognizable by their single nectar spur. The unusually colored red species in the reserve, red larkspur (*D. nudicaule*), has evolved a modified flower shape to attract hummingbird pollinators.

**Elegant brodiaea** (*Brodiaea elegans*). Elegant brodiaea is particularly lovely with its deep purple, funnel-shaped flowers. The bulbs of various species of brodiaea were one of the most significant sources of starch for some Sierran tribes. Native Americans gathered them with specially whittled and fired digging sticks. They ate the bulbs raw (some have a sweet, nutty flavor), roasted, fried, or boiled.

**Fiddleneck** (*Amsinkia menziesii*). Also known as rancher's fireweed, this plant generally grows in open, disturbed areas. The variety in the reserve has small orange flowers that grow along a coiled spike. This coil unrolls as the plant begins to blossom, and the curved appearance of the spike while unrolling resembles the neck of a fiddle. Native Americans pounded fiddleneck seeds into flour, formed them into cakes, and ate them uncooked.

**Foothill mule-ears** (*Wyethia helenioides*). Foothill mule-ears is one of the many wild sunflowers living in Cold Canyon, easily identified by its very large yellow flower head and broad, almost fuzzy, leaves.

**Fremont's star lily** or **death camas** (*Zigadenus fremontii*). Although many members of the Lily family, like brodiaea, are edible, some are deadly. Fremont's star lily (also called death camas), which grows in Cold Canyon, is highly toxic. There have been unfortunate instances where people attempting to collect edible bulbs have instead collected death camas, a potentially fatal mistake.

**Henderson's shooting star** (*Dodecatheon hendersonii*). In the early spring, shooting stars blanket the cool creek bottoms and dot the trailside in Cold Canyon. Dainty and strikingly magenta to lavender, the blossoms are impossible to overlook. Native Americans roasted stems and leaves of this plant for food, and used the blossoms as a poultice to reduce swelling.

**Ithuriel's spear** (*Triteleia laxa*). The lovely Ithuriel's spear has many bluish-purple flowers in a large, loose cluster. This plant is a close relative of elegant brodiaea (*Brodiaea elegans*), but unlike *Brodiaea elegans*, the petals of Ithuriel's spear are not reflexed (curved away from the flower's center).

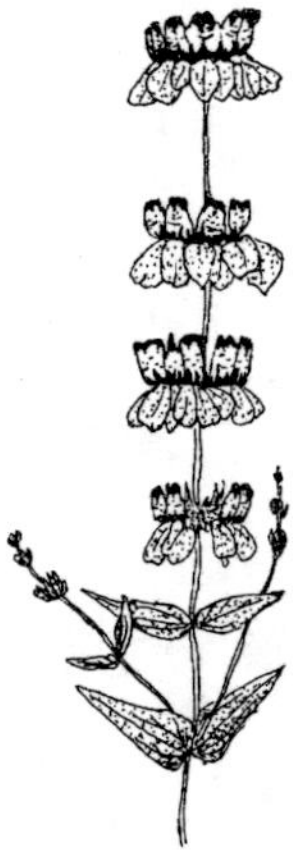

**Chinese houses**

**Larkspur**

**Elegant brodiaea**

**Fiddleneck**

**Foothill mule ears**

**Fremont's star lily**

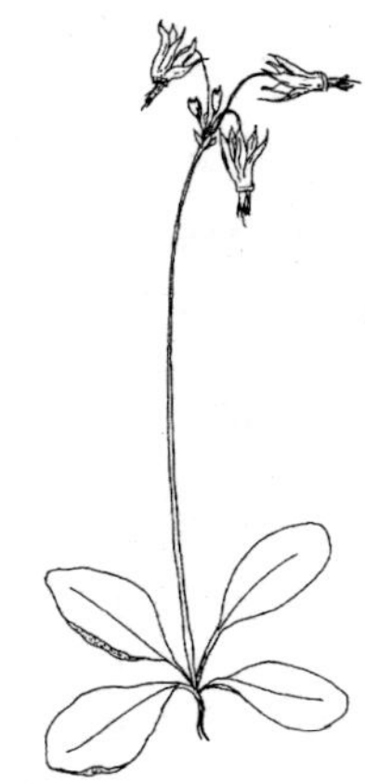

**Henderson's shooting star**

**Ithuriel's spear**

## VEGETATION

**Miner's lettuce** (*Claytonia perfoliata*). Miner's lettuce leaves are rounded and shiny, surrounding the stem, and the small flowers are white or pinkish. Sometimes the leaves or stem have a reddish tinge. The black, shiny seeds of this plant are an important food for many songbirds. The leaves are an excellent source of vitamin C.

**Miniature lupine** (*Lupinus bicolor*). This perennial can be found growing in open areas of the reserve. Lupines can be identified by their palm-like leaves, with seven to nine leaflets arranged in a fan. Their flowers, like all legumes, have a central "banner" and four smaller petals. The banner in the miniature lupine is turned upwards, and is usually white or yellow with blue dots, while the other petals are usually blue. After flowering in the spring, lupines produce the hairy pods typical of all legumes. Unlike other legumes, however, the seeds from lupines are poisonous. Nevertheless, Native Americans would use a tea made from boiled lupine leaves and seeds to help cure gastrointestinal ailments.

**Paintbrush** (*Castilleja* spp.). Because these paintbrush-shaped plants occur in lovely shades of red, orange, and even yellow, some Native American tribes told their children that the paintbrush was used to paint the sunset. The colorful "petals" of the paintbrush flowers are not really petals at all, but sepals (the green flower parts surrounding the petals in other flowers) and "bracts" (modified leaves surrounding the sepals). Paintbrushes often derive some of their nourishment by parasitizing the roots of other plants and are pollinated by hummingbirds.

**Soap plant** (*Chlorogalum pomeridianum*). Early settlers named this plant "soap plant" because they dug up its root, stripped it of the fibrous outer coating, and rubbed it on their hands underwater to make a lather. Soap plant has beautiful, white flowers of six petals that most people never see: the flowers are moth-pollinated and only open at night. The 1 inch (2 cm) wide, long, wavy leaves that sprout in a circle at the base of the plant is a sure sign of soap plant. Out of the center of this whorl often points a tall (over 3 ft. or 1 m) bare stalk with dozens of tightly closed, pill-shaped buds. Seasonally, these leaves are browsed to nearly ground level by deer and other mammals.

**Yarrow** (*Achillea millefolium*). Yarrow is a frequent resident of roadsides and other disturbed areas, easily recognizable by its flat-topped clusters of small, white flowers. The leaves are distinctively narrow and feathery, and have a strong but pleasant smell.

**Yellow star-thistle** (*Centaurea solstitialis*). One of California's most noxious weeds, yellow star-thistle made significant inroads into the reserve during the 1997-'98 El Nino year. It is toxic to livestock, and disperses via the wind and on the fur of mammals. As a result, star-thistle has spread rapidly in disturbed habitats in the Central Valley, the Coast Range, and the Sierra Nevada. In the reserve, it is most abundant in the field next to the old homestead, where the spines surrounding each flower are responsible for causing painful scratches on many hikers' legs. However, it is also prevalent on the trail, spread in part by the many boots traveling up and down the track. The ability of star-thistle to spread by catching a ride on the paws and backs of dogs is just one reason why dogs are not allowed on the reserve. Through concentrated management efforts, it may be eradicated from the reserve, but its presence nearby is a constant threat.

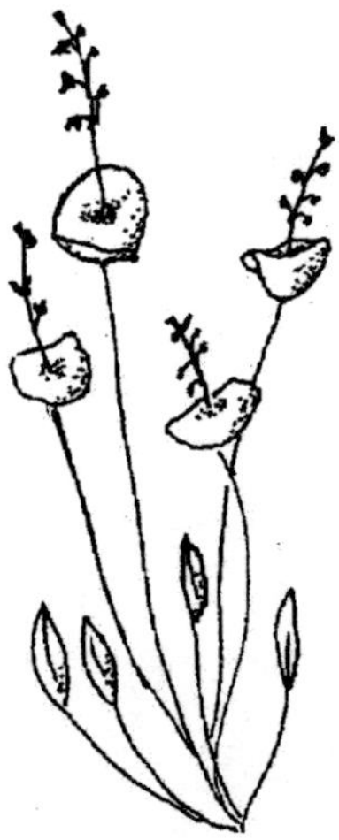

Miner's lettuce

Paintbrush

Miniature lupine

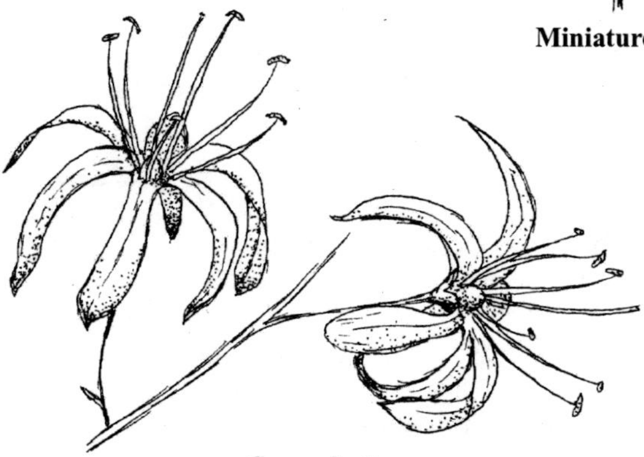

Soap plant

Yellow star-thistle

Yarrow

# Blooming Calendar of Cold Canyon's Flowering Plants

This blooming calendar indicates the month in which particular species commonly begin blooming. Climatic variation may alter the blooming calendar. For example, in colder years, the timing of blooming for some species may be delayed.

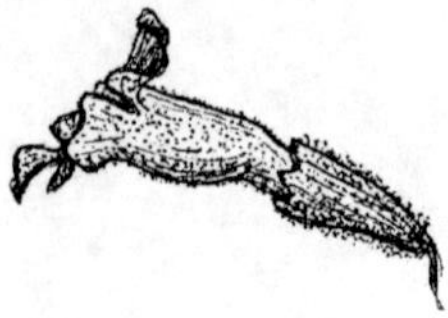

| Common Name | Scientific Name | Family |
|---|---|---|
| | **January** | |
| Milk maids | *Cardamine californica* | Brassicaceae |
| | **February** | |
| California pipevine | *Aristolochia californica* | Aristolochiaceae |
| Large-leaved hound's tongue | *Cynoglossum grande* | Boraginaceae |
| Wild cucumber | *Marah fabaceus* | Cucurbitaceae |
| Parry manzanita | *Arctostaphylos manzanita* | Ericaceae |
| Red maids | *Calandrinia ciliata* | Portulacaceae |
| Miner's lettuce | *Claytonia perfoliata* | Portulacaceae |
| Henderson's shooting star | *Dodecatheon hendersonii* | Primulaceae |
| California saxifrage | *Saxifraga californica* | Saxifragaceae |
| Indian warrior | *Pedicularis densiflora* | Scrophulariaceae |
| | **March** | |
| Gamble weed | *Sanicula crassicaulis* | Apiaceae |
| Fiddleneck | *Amsinckia menziesii* | Boraginaceae |
| Western redbud | *Cercis occidentalis* | Fabaceae |
| Fleshy lupine | *Lupinus succulentus* | Fabaceae |
| Vari-leaf nemophila | *Nemophila heterophylla* | Hydrophyllaceae |
| Caterpillar phacelia | *Phacelia cicutaria* | Hydrophyllaceae |
| Imbricate phacelia | *P. imbricata* | Hydrophyllaceae |
| Annual California poppy | *Eschscholtzia caespitosa* | Papaveraceae |
| Bird's eyes | *Gilia tricolor* | Polemoniaceae |
| True baby stars | *Linanthus bicolor* | Polemoniaceae |
| Virgin's bower | *Clematis lasiantha* | Ranunculaceae |
| Blue larkspur | *Delphinium hesperium* | Ranunculaceae |
| Red larkspur | *D. nudicaule* | Ranunculaceae |
| Jim brush | *Ceanothus oliganthus* | Rhamnaceae |
| Buck brush | *C. cuneatus* | Rhamnaceae |
| Sandbar willow | *Salix exigua* | Salicaceae |
| Hill star | *Lithophragma heterophyllum* | Saxifragaceae |
| Purple nightshade | *Solanum xantii* | Solanaceae |
| Blue dicks | *Dichelostemma capitatum* | Liliaceae |
| Checker-lily | *Fritillaria affinis* | Liliaceae |

| | **April** | |
|---|---|---|
| Yarrow | *Achillea millefolium* | Asteraceae |
| Foothill mule ears | *Wyethia helenioides* | Asteraceae |
| Western Wallflower | *Erysimum capitatum* | Brassicaceae |
| Blue elderberry | *Sambucus mexicana* | Caprifoliaceae |
| Pacific peavine | *Lathyrus vestitus* | Fabaceae |
| Bush lupine | *Lupinus albifrons* | Fabaceae |
| White-whorled lupine | *Lupinus microcarpus* | Fabaceae |
| Slender clover | *Trifolium gracilentum* | Fabaceae |
| Rose clover | *T. hirtum* | Fabaceae |
| Tomcat clover | *T. willdenovii* | Fabaceae |
| Toyon | *Heteromeles arbutifolia* | Rosaceae |
| Chinese houses | *Collinsia heterophylla* | Scrophulariaceae |
| Spinster's blue-eyed Mary | *C. sparsiflora* | Scrophulariaceae |
| Bush monkeyflower | *Mimulus auriantiacus* | Scrophulariaceae |
| California figwort | *Scrophularia californica* | Scrophulariaceae |
| Blue-eyed grass | *Sisyrhinchium bellum* | Iridaceae |
| Golden fairy lantern | *Calochortus amabilis* | Liliaceae |
| Ithuriel's spear | *Triteleia laxa* | Liliaceae |
| Fremont's star lily | *Zigadenus fremontii* | Liliaceae |
| | **May** | |
| Yellow star-thistle | *Centaurea solstitialis* | Asteraceae |
| Woolly sunflower | *Eriophyllum lanatum* | Asteraceae |
| Spice bush | *Calycanthus occidentalis* | Calycanthaceae |
| California broom | *Lotus scoparius* | Fabaceae |
| California tea | *Rupertia physodes* | Fabaceae |
| California buckeye | *Aesculus californica* | Hippocastanaceae |
| Yerba santa | *Eriodictyon californicum* | Hydrophyllaceae |
| Pitcher sage | *Lepechinia calycina* | Lamiacaeae |
| Red ribbons | *Clarkia concinna* | Onagraceae |
| Four-spot | *C. purpurea* | Onagraceae |
| Elegant clarkia | *C. unguiculata* | Onagraceae |
| Chamise | *Adenostoma fasciculatum* | Rosaceae |
| Valley tassels | *Castilleja attenuata* | Scrophulariaceae |
| Woolly paintbrush | *C. foliolosa* | Scrophulariaceae |
| Foothill penstemon | *Penstemon heterophyllus* | Scrophulariaceae |
| Long-spurred plectritis | *Plectritis ciliosa* | Valerianaceae |
| Long-tubed iris | *Iris macrosiphon* | Iridaceae |
| Elegant brodiaea | *Brodiaea elegans* | Liliaceae |
| Twining lily | *Dichelostemma volubile* | Liliaceae |
| | **June** | |
| Common madia | *Madia elegans* | Asteraceae |
| Whispering bells | *Emmenanthe penduliflora* | Hydrophyllaceae |
| Downy navarretia | *Navarretia pubescens* | Polemoniaceae |

## VEGETATION

| | **June (continued)** | |
|---|---|---|
| Many-flowered buckwheat | *Eriogonum dasyanthemum* | Polygonaceae |
| Martin's Indian paintbrush | *Castilleja applegatei* | Scrophulariaceae |
| California wild grape | *Vitis californica* | Vitaceae |
| Gold nuggets | *Calochortus luteus* | Liliaceae |
| | **July** | |
| California rock parsnip | *Lomatium californicum* | Apiaceae |
| Yampah | *Perideridia kellogii* | Apiaceae |
| Narrow-leaved milkweed | *Asclepias fascicularis* | Asclepiadaceae |
| Coyote brush | *Baccharis pilularis* | Asteraceae |
| Green cudweed | *Gnaphalium californicum* | Asteraceae |
| California sunflower | *Helianthus californicus* | Asteraceae |
| Leather root | *Hoita macrostachya* | Fabaceae |
| White sweet clover | *Melilotus alba* | Fabaceae |
| Klamathweed | *Hypericum perforatum* | Hypericaceae |
| Coyote mint | *Monardella villosa* | Lamiacaceae |
| | **August** | |
| Mugwort | *Artemisia douglasiana* | Asteraceae |
| Twiggy wreath plant | *Stephanomeria virgata* | Asteraceae |
| California fuchsia | *Epilobium canum* | Onagraceae |
| | **September** | |
| Broad-leaved aster | *Aster radulinus* | Asteraceae |
| California brickellia | *Brickellia californica* | Asteraceae |
| Small-flowered calycadenia | *Calycadenia pauciflora* | Asteraceae |
| Western goldenrod | *Euthamia occidentalis* | Asteraceae |
| Great Valley grindelia | *Grindelia camporum* | Asteraceae |
| Field tarweed | *Hemizonia congesta* | Asteraceae |
| Prickly lettuce | *Lactuca serriola* | Asteraceae |
| | **November** | |
| Chaparral currant | *Ribes malvaceum* | Grossulariaceae |

# Invertebrates of Cold Canyon

The words "animals of Cold Canyon" probably bring to mind deer, newts, hummingbirds and the like. However, most animals in Cold Canyon, measured in sheer numbers or in species diversity, are invertebrates: insects, spiders, slugs, worms, and other animals lacking backbones. In fact, invertebrates comprise over 95% of the millions of species inhabiting the earth. Many of these organisms live only in the ocean; however, a huge diversity of arthropods (invertebrates like insects and spiders) live in the reserve, and other invertebrates such as worms, slugs, and snails are present but less conspicuous. To put the levels of diversity into perspective, realize that the reserve holds no more than fifty mammal species but probably thousands of insect species and hundreds of spider species, most of which have not even been identified.

Invertebrates are often ignored or dismissed by the average nature lover, who is usually only concerned with the pesky buzzing of mosquitoes or the eerie tickle of a tick. Most of the invertebrates in Stebbins Cold Canyon Reserve, however, are harmless to humans and even entertain the keen observer with a wealth of interesting natural history. While hiking the trails or enjoying lunch by the creek, take note of the invertebrate world around you. Colorful butterflies may pause to "mud-puddle" on the moist soil of the streambed, where they sip essential minerals. Water striders will entertain you with their pulsing dance on the water's surface as dragonflies hover above, their iridescent bodies glinting in the sun. Leopard slugs move imperceptibly across fallen leaves and moss, leaving behind a trail of shimmering slime.

**Invertebrates like this leopard slug abound in Cold Canyon.**

## Habitats

Because invertebrates are so small compared to us, tiny changes in the environment that we barely notice may constitute completely different habitats for them. Our eyes readily perceive the canyon bottom as a single riparian plant community habitat. However, if we could observe the riparian community through a bug's eye, we would find a multitude of microhabitats determined by plant composition, relative moisture, soil type, sun exposure, or temperature (Table 1). Some invertebrates are generalists and range freely across all these microhabitats. Specialists, on the other hand, can be found only in one or two of these microhabitats and cannot survive elsewhere. For instance, wasps and dragonflies, which are generalist predators, can be found flying in many different communities. However, the valley elderberry longhorn beetle (*Desmocercus californicus dimorphus*), a specialist herbivore, spends the vast majority of its life inside the branches of one creekside elderberry bush (*Sambucus mexicana*). As observers of nature, we must realize that differ-

## INVERTEBRATES

**Table 1.** Microhabitats of common invertebrates found in the plant communities of Cold Canyon.

| Plant Community | Microhabitat | Common Invertebrates |
|---|---|---|
| Riparian | Pools | Backswimmers, waterboatmen, giant water bugs, dragonfly and damselfly nymphs |
| | Under stones | Stonefly and mayfly nymphs |
| | Rock surfaces | Black fly larvae |
| | Water surface | Water striders, whirligig beetles |
| | Emergent plants | Dragonflies and damselflies, crane flies, orb-weaving spiders |
| | Streamside | Toad bugs, slugs |
| Live oak woodland, Savanna , Chaparral, Grassland | Under ground | Ants, bees, wasps, earthworms |
| | Ground surface | Antlions, grasshoppers, beetles |
| | Leaf litter, under rocks | Earwigs, predaceous ground beetles, snails, slugs, scorpions, ground spiders, millipedes |
| | Rotting logs | Carpenter ants, termites, pill bugs |
| | Low-growing plants | Aphids, stink bugs, butterfly larvae, ticks, spiders |
| | In woody vegetation | Valley elderberry longhorn beetles, bark beetles |
| | Tree canopy | Moths, aphids, lacewings, mites |
| | Flowers | Butterflies, bees, bee flies, ants, beetles, crab spiders |
| | Galls | Wasps, moths |

ent animals view the world at different spatial scales: a puddle to a coyote may be a water strider's entire universe.

In your quest to discover the invertebrates of Cold Canyon, keep in mind that they thrive in both aquatic and terrestrial habitats. As you hike up the canyon, linger by the creek and look for invertebrate life. The stream bank is itself an active invertebrate habitat. Large slugs move lazily through the leaf litter; spiders patrol the shore in search of prey; and a multitude of worms and nematodes lurk just below the surface. On, in, and under the water, riparian and aquatic invertebrates thrive. Surprisingly, the greatest diversity of aquatic insects is often hidden in the debris and rocks on the bottom of the creek. Above them on the creek's surface, some insects have the remarkable ability to walk on water. Sitting next to a pool in springtime will allow you to observe numerous insects as they glide, buzz, and whirl in an aesthetically pleasing but seemingly purposeless dance. In actuality these highly adapted insects are actively hunting or courting as they swirl across the surface. Look for other insects swimming about in calm, open water or clinging to aquatic vegetation. Spider webs, mollusk shells, and hollow insect pupal cases can often be found still attached to a bit of vegetation or a rock, evidence of the ubiquitous presence of invertebrates.

As you move away from the creek, look for invertebrates on different parts of the woodland trees and shrubs. A single oak is like a thriving arthropod metropolis; beetles, moths, and spiders hide in its bark, while other insects live off the canopy's leaves. The grasses and flowering plants will harbor yet a different set of invertebrates. As you look at a flower, notice that some insects eat the leaves or stems of the plant, other insects only sip nectar, and certain spiders have evolved to hunch motionless on the plant, waiting for these passersby. Also, do not miss the activity right beneath your feet. Cold Canyon's worms and many of its insects can be found in subterranean burrows. Ants create underground colonies, which are sometimes parasitized by other insects.

## Taxonomy

Taxonomists classify each organism by placing it into a kingdom, phylum, class, order, family, genus, and species. These groups are hierarchically organized, so related species share a common genus, related genera (plural of "genus") share a common family, related families share a common order, and so forth. The accompanying evolutionary tree on page 42 shows the relationships among the most common invertebrate groups as they are currently understood by biologists. Groups connected by a single branching point share a common ancestor, and are thus more closely related than phyla separated by several branching points. For instance, the phyla Mollusca, Annelida, and Arthropoda are each more closely related to each other than they are to the phylum Chordata (the phylum that includes mammals and other vertebrates). Likewise, the phylum most closely related to Chordata is Echinodermata.

The evolutionary tree also indicates that the term "invertebrates" is not a rigorous classification because it refers to a hodgepodge of animals whose only commonality is that they do not belong to our group, the subphylum Vertebrata. Nevertheless, we divide Cold Canyon's fauna into invertebrates and vertebrates (in the next chapter) for several reasons. First, people can usually distinguish between a vertebrate and an invertebrate animal, even if they have trouble distinguishing the different subgroups. Second, although people can usually tell the difference between two closely related vertebrate species, such as bald and golden eagles, the same is often not true for invertebrates: some related species cannot be differentiated without a microscope. Fortunately, many invertebrates share characteristics at higher taxonomic levels, such as at the level of the family or order. For example, distinguishing between Lepidoptera (the order that comprises the butterflies and moths) and Odonata (the order including dragonflies and damselflies) is relatively easy. Third, the sheer diversity of the invertebrate world defies the scope of this field guide. For example, the order Lepidoptera is represented by at least 500 species of moths and butterflies in the reserve. Thus, in Cold Canyon, that single order has ten times the diversity of the mammals, and a guide that offered descriptions of every worm, spider, insect, and mollusk that occurred in the reserve would be far too heavy to take along on a hike! Finally, the division highlights the fact that all vertebrates of Cold Canyon have been identified to species level, but many fewer invertebrates in Cold Canyon have achieved this level of identification. Hence, the distinction between vertebrates and invertebrates indicates that we simply do not know much about the species diversity of invertebrates.

Nevertheless, the information in this chapter will help you to identify many of the reserve's invertebrates. Table 2 provides a list of the main invertebrate groups covered in this field guide. We focus primarily on mollusks, spiders, and insects, and cover other groups only briefly. In most cases, we identify distinctions only among the major groups of organisms, i.e., at the phylum and order levels. However, some families, genera, and

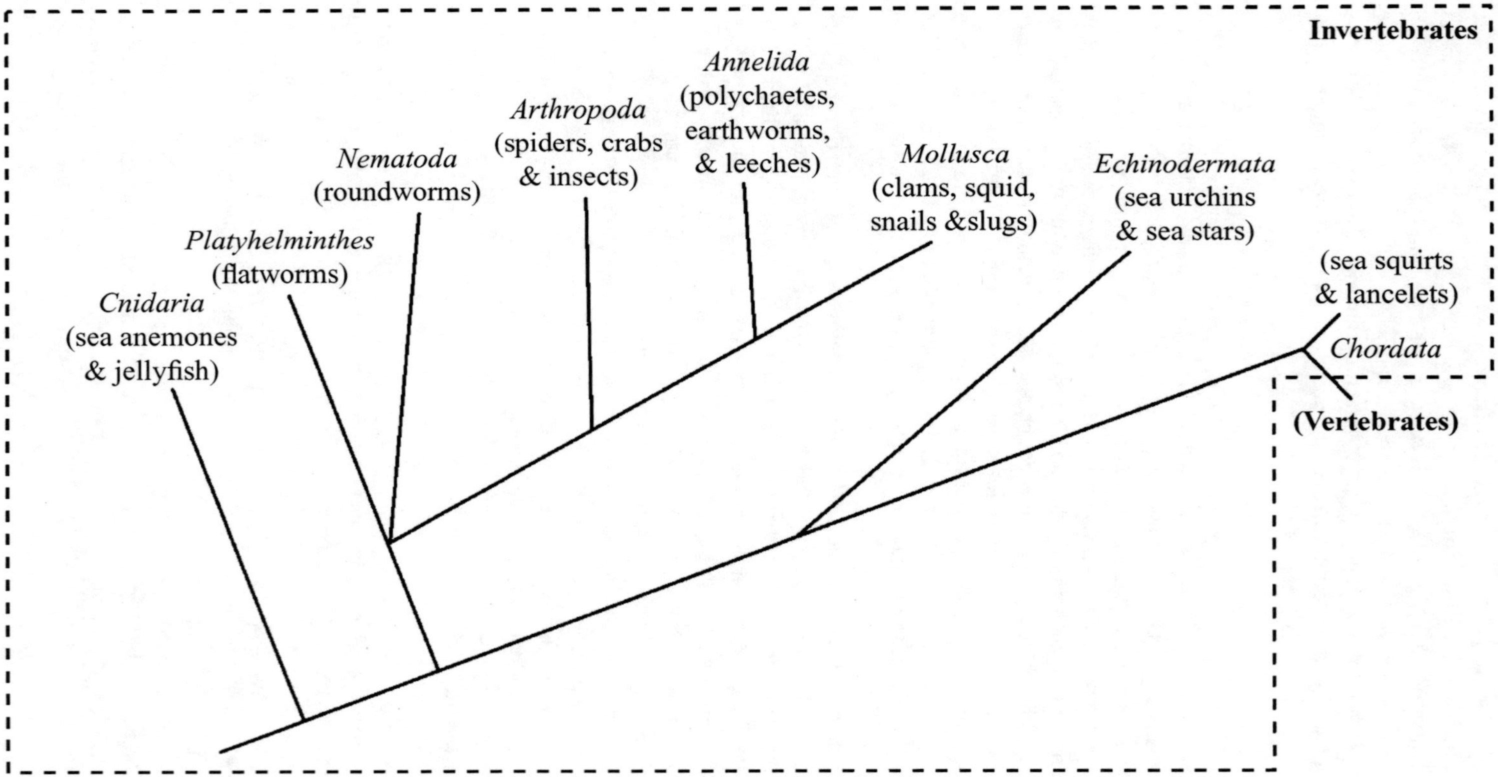


**The major phyla of the animal kingdom. As indicated by the dotted line, most phyla are invertebrates, and vertebrates are only a part of the phylum Chordata.**

**Table 2.** The major invertebrate phyla and classes of Cold Canyon.

| PHYLUM/Subphylum | Class |
|---|---|
| ANNELIDA | Oligochaeta (earthworms) |
| MOLLUSCA | Gastropoda (slugs and snails) |
| ARTHROPODA | |
| Chelicerata | Arachnida (arachnids) |
| Crustacea | Malacostraca (amphipods, sowbugs, and pill bugs) |
| Uniramia | Diplopoda (millipedes)<br>Chilopoda (centipedes)<br>Insecta (insects) |

species can be readily identified and are covered individually in the text. For example, the black and iridescent-green wings of the pipevine swallowtail (Papilionidae: *Battus philenor*) differentiate it from all other butterfly species in the reserve. In such cases, this field guide provides information to allow observers to spot these particular species.

## Annelids (Phylum Annelida)

Annelids are segmented worms and include polychaetes (class Polychaeta), leeches (class Hirudinea), and earthworms (class Oligochaeta). Polychaetes are strictly marine and free-living; leeches are aquatic and usually parasitic; and earthworms are free-living in moist detritus or soil. Cold Canyon obviously does not support polychaetes and probably does not have leeches because of a lack of their food sources.

**Earthworms** (Class Oligochaeta). Earthworms are tubular, segmented, soft-bodied animals that subsist on organic material in the soil they ingest. They are perhaps most associated in the human mind with aerating garden soil; however, earthworms actually perform this same service all over the globe in natural systems as well. Although they have not yet been surveyed, earthworms can undoubtedly be found in Cold Canyon. They are probably easiest to find when they emerge at the soil surface after hard rains.

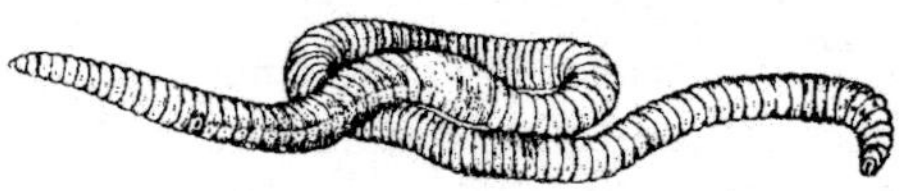

**Earthworm**

## Mollusks (Phylum Mollusca)

Mollusks are the group that includes octopuses, squid, clams, oysters, and snails. Although these animals are all quite diverse in body form and lifestyle, they have certain commonalities that betray their shared evolutionary history. For instance, they all have a radula, a tongue-like organ lined with sharp teeth-like protrusions, used for feeding. Mollusks are a successful group in evolutionary terms, which means that they have evolved into many different species. All mollusks share a soft, unsegmented body, and their skin is not covered by a chitinous external skeleton as in insects or a cuticle like our own skin. This means that they cannot control evaporation of water from their skins, a limitation which restricts most kinds of mollusks to an aquatic existence. Because they are vulnerable in dry environments, even terrestrial mollusks inhabit only moister areas.

The only mollusks that occur in Cold Canyon are snails and slugs. When most people think of snails, they think of the common garden snail, which eats holes in their petunias and young tomato plants and speckle sidewalks and garden paths at night. In fact, the brown gardensnail (*Helix aspersa*) is only one of an estimated 30,000 kinds of terrestrial snails worldwide and is unusual in its habit of feeding on green, living plants. Most snails are the size of a pebble and never make themselves known to gardeners and farmers. Instead, they live in moist, shady places, quietly feeding on dead and decaying plant material, thus reducing the detritus to ever-smaller particles, which speeds up decomposition and soil formation.

Snails have shells made out of calcium carbonate, into which their internal organs are coiled, and slugs are their somewhat tubular relatives that through evolution have ceased to have a shell. A small evolutionary remnant of the ancestral shell (a "vestigial" shell) sometimes occurs buried in the tissues of the slug's back. There are several hundred kinds of slugs worldwide, but they are mostly restricted to cool, humid, and generally forested regions.

Snails and slugs are hermaphroditic, which means that each individual can produce sperm and eggs at the same time. Nevertheless, to produce offspring, they usually seek to mate with other individuals of their species. Mating is often preceded by a lengthy "courtship" of mutual circling and touching, which scientists believe sends a message to its partner that it can produce healthy offspring, much like the message male birds send with their bright plumage.

Cold Canyon Reserve has not been studied by malacologists (mollusk biologists), so we do not know exactly what kinds live here. You may actually be the first person to discover that a species occurs in the reserve, and interested malacologists are encouraged to study this area. The following information is based on the distributions of snails and slugs elsewhere in the Coast Ranges of California.

**Shoulderband** (*Helminthoglypta* spp.). The shoulderband, a snail with a shell about 1 inch (2 cm) in diameter, may be a reserve inhabitant. Shoulderbands look similar to brown gardensnails (*Helix aspersa*) known from human-dominated California landscapes, but their tan shells have a single narrow dark band rather than a broad zone of dark, cloudy pigment. The name shoulderband refers to that narrow band on the "shoulder" of the shell. More than 60 species of *Helminthoglypta* are native to California.

**Shoulderband shell**

**Lancetooth snail** (*Haplotrema* spp.). Lancetooth snails are flat-shelled snails without bands; the shell is greenish-yellow with a waxy luster and is about 0.5 to 0.75 inches (1-2 cm) in diameter. The head and foot of the animal are pale buff or cream-colored. *Haplotrema* is unusual among snails in that its diet also includes other snails; the

**Lancetooth snail**

"teeth" of its radula are sharp and elongate (leading to the common name "lancetooth"), presumably as an adaptation to this food source.

**Pacific banana slug or leopard slug** (*Ariolimax columbianus*). The only species of land mollusk already observed to occur in the reserve is a large yellowish or olive-brown slug, sometimes almost covered with black patches. This peculiar appearance explains its common names. Sometimes reaching almost 10 inches (25 cm) in length, the Pacific bananaslug is one of the largest species of slugs in the world. Most individuals in Cold Canyon Reserve are somewhat smaller. It has been suggested that the metabolic cost of producing mucus (which slugs and snails need to glide on) makes slugs much larger than this species biologically improbable.

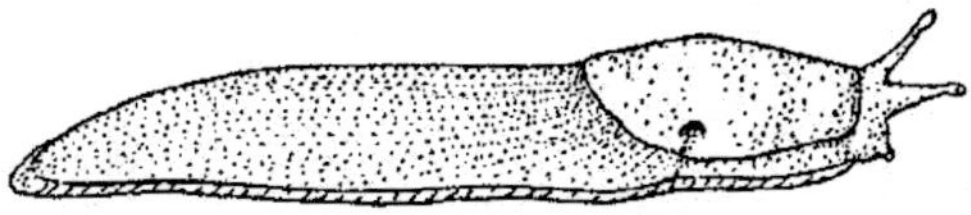

**Banana (or leopard) slug**

## Arthropods (Phylum Arthropoda)

The diversity of arthropods worldwide can only be guessed, but it is known to be enormous. The importance of arthropods in ecosystems is, in many cases, not well understood; however, ecologists believe that they often play pivotal roles in ecosystem processes such as nutrient recycling. Although arthropods are quite diverse in appearance, ranging from lobsters to millipedes to spiders to butterflies, they all have certain characteristics in common. For example, "arthropod" means "jointed foot," which hints at their jointed appendages. Also, all arthropods have tagmata, which are specialized body segments, such as the head, thorax, and abdomen in insects.

### *Chelicerata: Arachnids*

The class Arachnida, which includes the spiders (order Araneae), harvestmen (order Opiliones), mites and ticks (order Acari), and scorpions (order Scorpiones), has long attracted the attention of naturalists for its amazing diversity in morphology, ecology, and life history. Arachnids have conquered most of the earth's terrestrial and marine environments and play major roles as predators, herbivores, parasites, and decomposers in ecosystems worldwide. With more than 108 families and 40,000 species, spiders are the second largest order of arachnids and exhibit some of the most interesting biology known among invertebrate animals. Only in the insects can another assemblage of such diverse organisms be found.

Arachnids can be differentiated from other small, hard-bodied creatures in the reserve by their numbers of legs and antennae. Arachnids have four pairs of walking legs and no antennae, whereas insects have three pairs of walking legs and one pair of antennae. Many arachnids also have the startling characteristic of multiple pairs of eyes, which may be one reason for their success as predators.

Because of its numerous microhabitat types, Cold Canyon provides an excellent representation of Californian arachnid diversity. Most of the major orders occur in the reserve, including spiders, scorpions, mites and ticks, harvestmen, pseudoscorpions, and windscorpions. A current list of recorded families and habitats can be found in appendices (p. 108). To date only the spiders have been surveyed, leaving significant gaps in the understanding of what should prove to be a remarkable fauna. Arachnid researchers are encouraged to expand on this work through additional surveys or taxonomic work.

## INVERTEBRATES

### Mites and ticks (Order Acari)

Mites and ticks are both small, flattened, hard-bodied animals that occur in multiple habitats on the reserve. Mites are minute, almost microscopic creatures that may either feed on plants, other insects, mammals, or birds, depending on the species. They are ubiquitous and abundant, and can often be spotted by looking for red dots on plants in the reserve. Ticks, on the other hand, are not as common as mites, but because they have the unpopular habit of feeding on blood, they attract more attention. After the adult tick has fed on enough blood to nourish her eggs, she drops off her host and lays her eggs in the ground litter. After hatching from the eggs, her offspring work their way up a leaf of grass and wait to hitch a ride on the next passerby. Because the bite of a tick is not detectable, it is important to carefully examine your skin for ticks after hiking anywhere in California.

### Scorpions (Order Scorpiones)

Scorpions occur throughout the Coast Range and are an exciting if somewhat startling sight. They have stout bodies, large pincers, and a tail that curves over its body to deliver a poisonous sting to its prey. Their stings, though painful, are not toxic enough to be life-threatening to human beings. Scorpions usually lurk in dark places, so they should be sought under logs or rocks.

### Spiders (Order Araneae)

California is a center of spider diversity and endemism with 55 families and over 1,500 species represented. Despite its small area, Cold Canyon is home to an exciting amount of this diversity (20 recorded spider families, see Appendix 3, p. 107). This diversity is partly due to the wide array of available microhabitats in Cold Canyon. The following groups of spiders are some of the more common families found in the reserve. Where possible, the descriptions that follow emphasize web construction because webs are often the most straightforward means of identifying spiders.

**Funnel-web spiders** (Family Agelenidae). These spiders construct funnel-shaped webs leading into a shallow retreat near the ground or into dense vegetation. The funnels of grass spiders (*Agelenopsis* spp.) are found along the trail during spring, especially in tall, grassy vegetation. Grass spiders tend to be yellowish to brownish in color and have eight eyes. *Hololena* is among the most common spiders in the reserve and occurs in virtually every habitat although it is less abundant in riparian areas. In addition to the typical funnel, *Hololena* constructs an adjoining sheet that serves to signal the spider of approaching prey.

**Orb-weaving spiders** (Family Araneidae). Anyone who remembers Charlotte from childhood will instantly know the classic shape of an orb-weaver's web. The two genera with the largest body sizes in this family (0.8-1.2 in., or 2-3 cm, body length) in the reserve are *Araneus* and *Argiope*. Both genera consist of species that tend to be yellow to brown, but members of the genus *Argiope* are covered with silvery hairs. These spiders are abundant in the riparian areas of the reserve and are generally observed in their webs at dawn or dusk. In late summer their young are responsible for gossamer – the tangled strings of web often seen floating through the air. Gossamer results from ballooning juvenile spiders that abandon their mode of transport upon finding suitable new habitat to colonize.

**Tick**

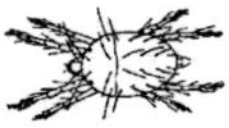

**Red spider mite**

**Scorpion**

**Funnel-web spider and web**

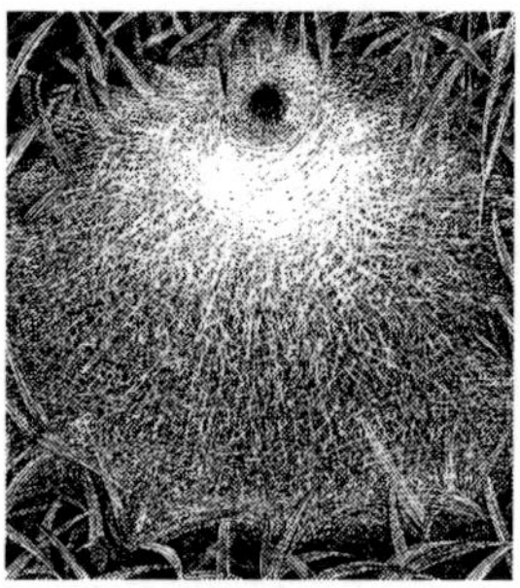

**Orb-weaving spider (*Araneus*) and web**

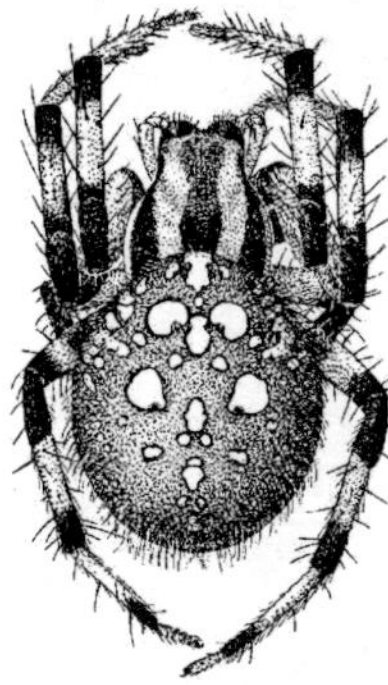

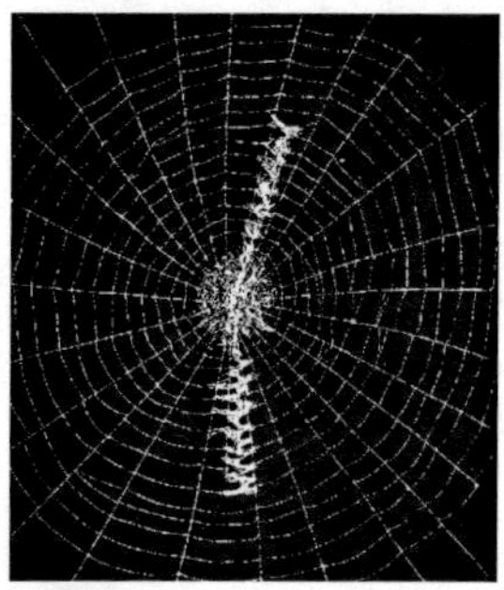

**Orb-weaving spider (*Argiope*) and web**

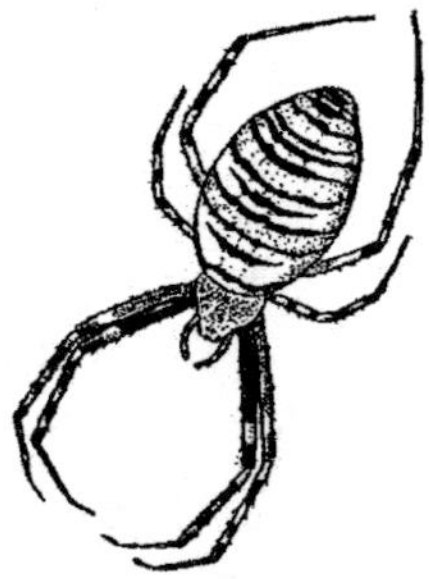

**Ground spider (*Micaria*)**

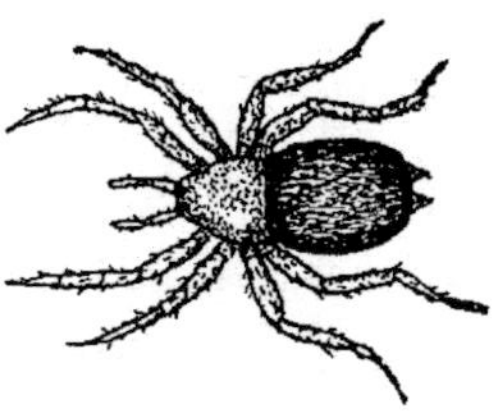

**Ground spider (*Sergiolus*)**

**Wolf spider (*Pardosa*), carrying offspring**

Funnel web spider web, social spider, black widow, and all webs on page 49 are reproduced with permission from McGraw-Hill Companies from Kaston and Kaston (1953).

## INVERTEBRATES

**Ground spiders** (Family Gnaphosidae). Ground spiders are among the most diverse spider families in California. They can be found in most microhabitats on the ground – under rotting logs, stones, or leaf litter. Most members of this family are dark and uniformly colored (not mottled like many other spiders). *Sergiolus montanus* is found under stones in the reserve, especially in cool areas. It has a black and white pattern on its abdomen and a distinctive "jerky" motion when disturbed. *Micaria* is found in the drier areas of the reserve and a constriction in its abdomen makes it an exceptional ant mimic.

**Wolf spiders** (Family Lycosidae). Wolf spiders are nocturnal predators and are notable for their pairs of eyes of unequal sizes. They are best observed at night (not on the reserve!) with a headlamp because their eyes reflect light and produce a distinctive iridescence. In most genera, the female carries the egg sac, and when the young emerge, they hitchhike on her abdomen for a considerable length of time. Two genera of wolf spiders occur in the reserve and are identified by size differences. The smaller (0.4-0.8 in., or 1-2 cm, body length) and very abundant thin-legged wolf spiders (*Pardosa* spp.) are common in riparian areas where they can be seen running along rock surfaces searching for prey. They acquired their name because of their relatively long, thin legs. The much larger and generally uncommon *Schizocosa* is most likely to be found among leaf litter in moist areas, though it may be difficult to spot due to its excellent mottled camouflage.

**Social spiders** (Family Pholcidae). Social spiders are one of the most common spiders at home – there are probably several living in the corners of your bedroom. Because they have small bodies and long legs, social spiders are often mistaken for "daddy longlegs" (also called harvestmen, Order Opiliones), which are not actually true spiders. Like many other spiders but unlike harvestmen, social spiders build large webs to trap prey. However, social spiders sometimes share webs with their neighbors. If you watch them closely, you can often see them communicating with each other by tapping on the lines of web.

**Jumping spiders** (Family Salticidae). Jumping spiders have the best vision of all spiders and can be observed in the open, slowly stalking their prey and suddenly pouncing. Their brightly colored appearance and complex behavior have captured the hearts of many would-be arachnophobes. *Habronattus americanus* is arguably the most beautiful of California's spiders and can be found in exposed sunny areas. Only the males of this species exhibit the wonderful contrast of red, white, and iridescent blue colors which have made this species famous. The larger *Phidippus audax* is the most abundant jumping spider in the reserve and is easily recognized by the white spots on its abdomen, its iridescent blue-green chelicerae, or biting mouthparts, and its extreme hairiness.

**Long-jawed orb weavers** (Family Tetragnathidae). Tetragnathid spiders are easily located by the small, horizontal (not vertical as in Araneidae) orb webs that are constructed over shallow streams or pools. These spiders have long abdomens, and bizarre, protracted jaws. When disturbed, these spiders align their legs vertically in the center of their webs, effectively mimicking twigs or other vegetation.

**Comb-footed spiders and widow spiders** (Family Theridiidae). Theridiid spiders are a large and disparate family of spiders much maligned due to the poisonous widow spiders

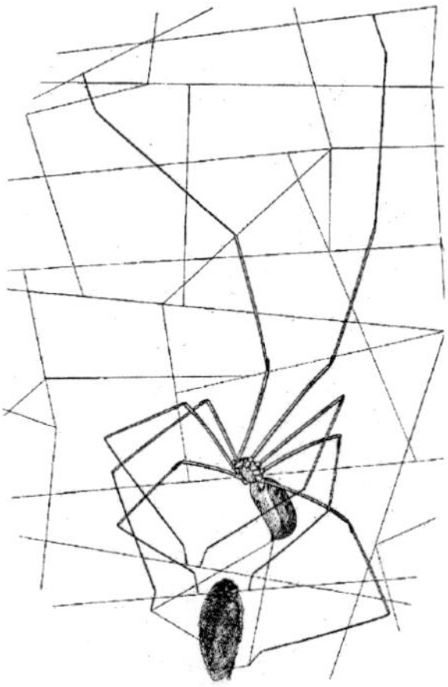

Social spider with prey

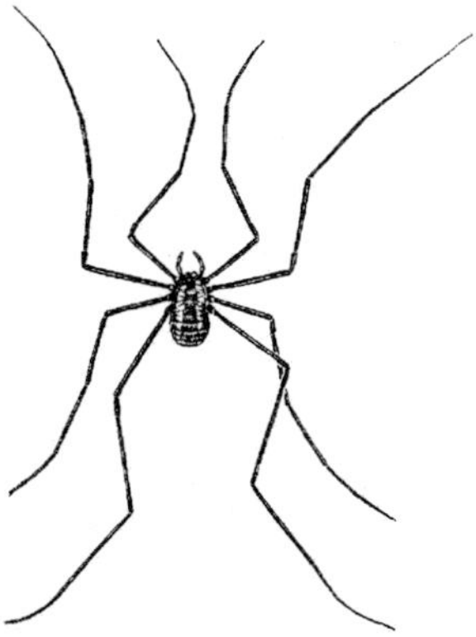

Harvestman (daddy longlegs)

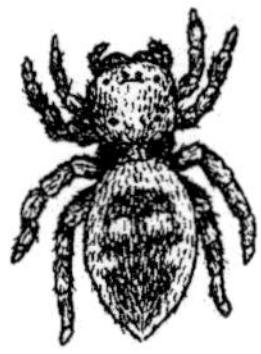

Jumping spider

Crab spider

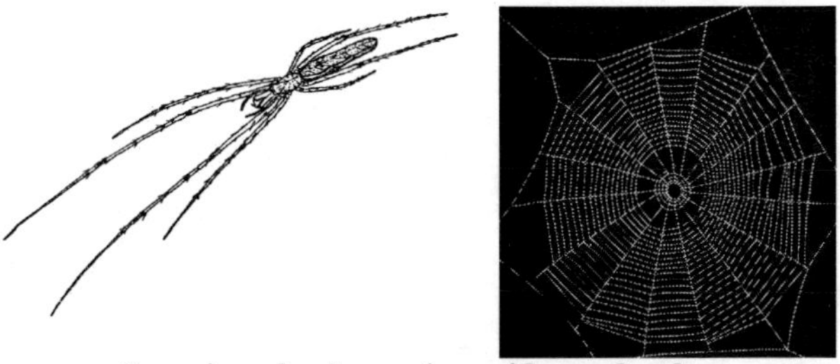

Long-jawed orb-weaving spider and web

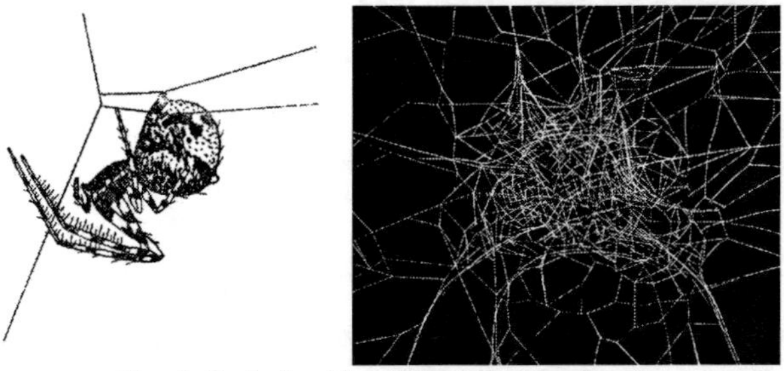

Comb-footed spider (*Theridion*) and web

Black widow

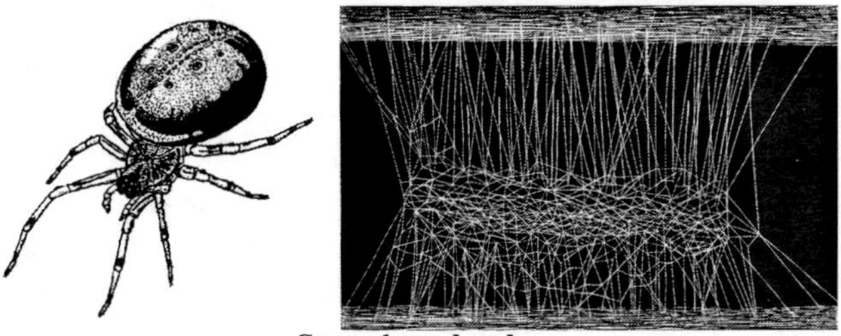

*Steatoda* and web

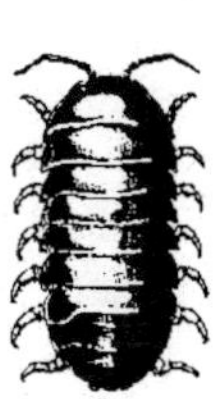

Pill bug

(*Latrodectus mactans* in California). While *L. mactans* does occur in the reserve, *Theridion* and *Steatoda* are more commonly encountered. *Theridion* is a very small spider (less than 0.2 in., or 5 mm, body length) of variable coloration, which builds irregular webs on vegetation and in crevices. *Steatoda* very closely resembles the black widow in size and shape, although it has no ventral hourglass marking and instead has a diagnostic white U on the back of the abdomen. Both *Steatoda* and black widows spin irregular webs and live in dark crevices, so it is critical to pay attention to detail when trying to distinguish between these two genera!

**Crab spiders** (Family Thomisidae). Crab spiders, which do indeed look like miniature crabs, are common throughout the reserve but are most abundant among trailside vegetation and riparian areas. They are sit-and-wait predators that build no webs. *Coriarachne* sports amazing bark-like camouflage that is so effective that the inexperienced or impatient eye will often not see this spider. *Coriarachne* is common on the bark of oak trees, especially at the edges of riparian areas.

## *Crustaceans*

Crustaceans, including pill bugs, sowbugs, crayfish, and crabs, have two pairs of antennae and variable numbers of legs. Most crustaceans are aquatic, and therefore cannot live in Cold Canyon when the creek dries up. The only crustaceans likely to be found are therefore pill bugs and sowbugs, and these are still restricted to moist microhabitats.

**Pill bugs and sowbugs** (Class Isopoda). Also called "rolly-pollies," pill bugs have the distinctive ability to roll up into a perfect ball when disturbed, and can be found under most rocks or logs where there is abundant detritus, their primary food source. Sowbugs look quite similar to pill bugs, but are flatter and cannot roll into a tight ball.

## *Millipedes, Centipedes, and Insects*

### Millipedes and centipedes (Class Diplopoda and Chilopoda)

When presented with a millipede, a centipede, and an insect, most people would be able to identify the insect right away, but they might stumble over the other two. However, a few simple points clearly distinguish them. First, millipedes are rounded like a finger, and centipedes are flattened. Second, millipedes have two pairs of legs per segment, while centipedes have only one pair per segment. It is important to make these distinctions before deciding to pick one up – millipedes eat detritus and are harmless to humans, but centipedes are predators and can deliver a nasty bite.

### Insects (Class Insecta)

At first glance, insects may appear to be a uniform group of small, drab creatures. Fortunately, a closer look reveals much more exciting news. They are an extraordinarily diverse group, and their role in ecosystems is fascinating. Because insects are critical players in a number of ecological processes, E. O. Wilson, one of the most famous ecologists of our time, calls insects "the little things that run the world." For example, many terrestrial insects are responsible for breaking down fallen leaves and dead wood into fertile soil, and aquatic insects provide food for multitudes of other insects, spiders, and birds.

Insects are well-known for their diversity of form at different life stages, and their ability to undergo metamorphosis. Metamorphosis can be incomplete or complete. In incomplete metamorphosis, young insects (often called nymphs) transform into similar-looking adults except that they acquire wings and sexual organs. Insects exhibiting incomplete metamorphosis include dragonflies, grasshoppers, and termites. In complete metamorphosis, the larvae of insects transform into very different-looking adults. Butterflies, beetles, ants, and bees fall into this category. Many larvae of thse insects are roughly the shape of stuffed sausages and no more attractive, but others have an appeal similar to caterpillars.

Because it takes a great deal of metabolic energy to undergo metamorphosis, inquisitive natural historians have long wondered why insects would bother. Unlike humans and other vertebrates, whose skeleton is on the inside, insects wear their skeleton on the outside. They molt, or shed their skeleton, to grow larger. This process allows them the flexibility to change form completely in a way the vertebrate internal skeleton does not (except in the case of amphibians). In addition, metamorphosis – especially complete metamorphosis – allows insects to avoid competition with their offspring for food. For instance, caterpillars eat foliage, but adult butterflies sip nectar.

Taxonomic experts at the University of California, Davis, have as yet cataloged only the insect orders and families of the reserve (see appendix on page 108 for a complete list of families). The following descriptions are grouped by insect orders (with the most ancestral orders appearing first), and some descriptions of particular families, genera, or species are also included. The taxonomy used in this guide comes primarily from *An Introduction to the Study of Insects* (Borror et al. 1989). The evolutionary relationships of many insects are still under debate, so be aware that the taxonomy in your insect field guide may vary somewhat from that used below. An indispensable resource for the interested entomologist is *California Insects* (Powell and Hogue 1979), which provides descriptions, interesting anecdotes, and illustrations of most of the insect species mentioned as well as additional ones not covered in this guide.

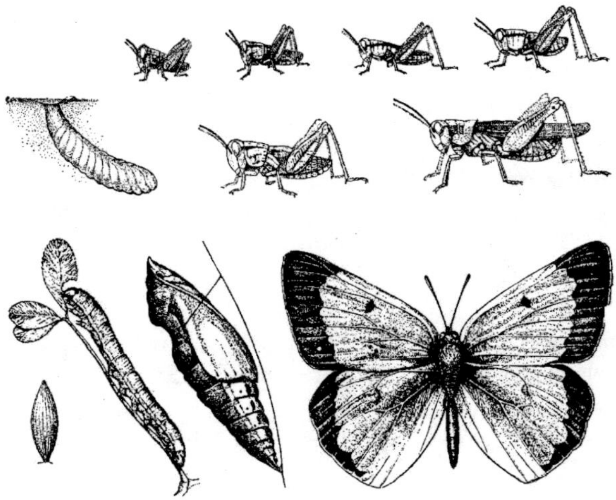

**Incomplete metamorphosis**

**Complete metamorphosis**

**The two main types of metamorphosis. Incomplete metamorphosis occurs in species like grasshoppers, and complete metamorphosis occurs in species like butterflies.**

## INVERTEBRATES

**Mayflies** (Order Ephemeroptera). The juveniles of this order are exceedingly diverse in form, but they are all aquatic and have a series of flap-like gills along each side of their abdomen. The adults have delicate bodies, translucent wings, and three long tails. Mayfly nymphs live on the stream bottom under rocks, sand, and submerged leaves (especially in riffles), but despite their inconspicuousness they are one of Cold Canyon's most abundant invertebrates. In their hidden world, insect nymphs feed and grow. Some filter feed, others wait surreptitiously for prey. Mayflies (Baetidae: *Baetis* spp. and Siphlonuridae: *Ameletus* spp.) are abundant, and, as herbivores, serve as an important source of food for both vertebrate and invertebrate aquatic predators. Although the juveniles spend months under the debris of the creek, the winged adult forms, which have no mouthparts and consequently cannot feed, spend only hours flying around in a mating dance above the creek.

**Dragonflies and damselflies** (Order Odonata). These familiar insects have long, thin bodies and two sets of elongated wings. Dragonflies can usually be found hovering in midair or at rest with their wings stiffly outstretched, whereas damselflies are most often seen perching, with wings folded above. These conspicuous and often beautiful insects congregate in clearings adjacent to the creek and the main trail, where swarms of gnats, other small flies, and mosquitoes provide the dragonflies and damselflies with ample food. The nymphs are also predatory, feeding on aquatic insects and, in some cases, even tadpoles or small fish. They are difficult to spot because of camouflage. They may either be a dull color or be covered with hairs that catch mud and help them to blend with the background. However, the adults are easy to spot. The common green darner (Aeshnidae: *Anax junius*) is a large bright green dragonfly with yellowish, transparent wings which buzzes around the ponds that form in the upper creek. A related dragonfly (Aeshnidae: *Aeshna multicolor*) sports two yellow diagonal markings on its dark thorax. Both are spectacular to watch as they perform their courtship and territorial acrobatics. Another common odonate in Cold Canyon is the blue damselfly (Coenagrionidae), which looks like a small, delicate dragonfly.

**Grasshoppers, crickets, and katydids** (Order Orthoptera). Members of this group are often recognized by their heavy bodies, mouthparts adapted for chewing plants, and the ability to jump. Despite this habit, many also have large wings. Jerusalem crickets (Stenopelmatidae: *Stenopelmatus fuscus*), which despite their name are native to this area, lurk in dark, moist places, such as under rocks and fallen logs. They may be up to 3 inches (8 cm) long, almost cartoon-like in their bumbling largeness. Recently, entomologists have discovered that these fascinating creatures exhibit species-specific ground-drumming patterns. The male plays a rhythmic serenade for the female, who responds only to the song of her own species. Some orthopterans communicate visually. As the summer approaches, sometimes the only activity visible to the hiker is that of grasshoppers (Acrididae: *Trimerotropis* spp. and *Arphia* spp.), which flash their yellow or orange hind wings as they fly about, seemingly oblivious to the hot, dry conditions around them. Other common grasshoppers on the reserve (Acrididae: *Melanoplus* spp.) have colorless hind wings.

**Earwigs** (Order Dermaptera). Every gardener is familiar with the earwig's general appearance – a dark elongated body with rear pincers. There is only one native species of earwig in California and its range does not extend to northern California. Nonetheless, a represen-

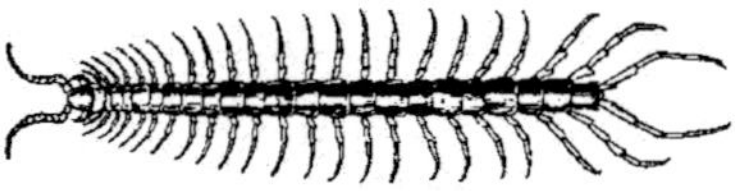

Centipede

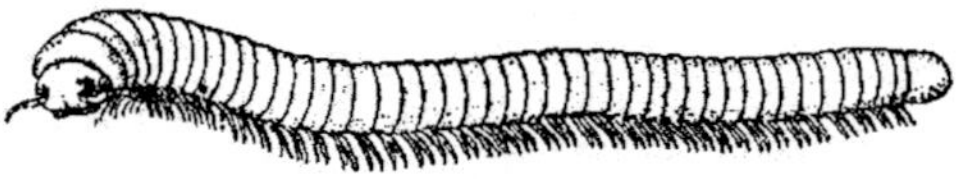

Millipede

Mayfly with nymph

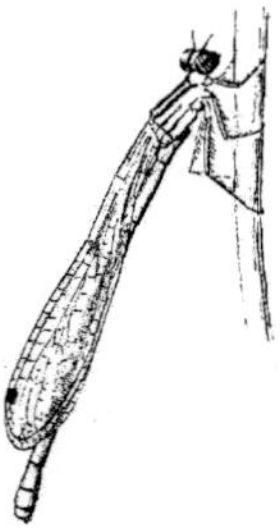

Damselfly

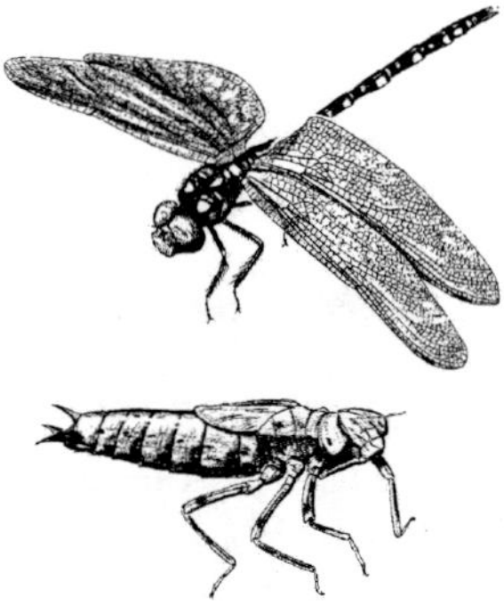

Dragonfly and nymph

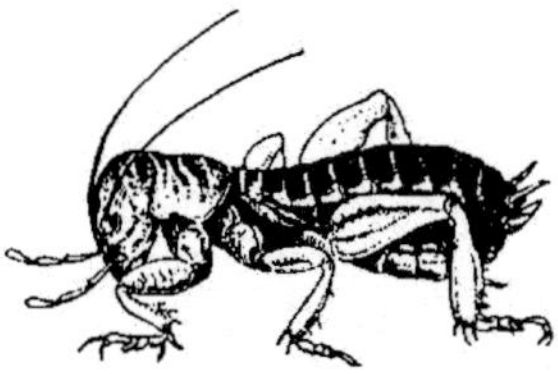

Jerusalem cricket

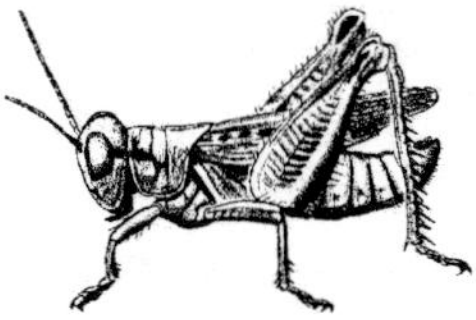

Grasshopper

Earwig

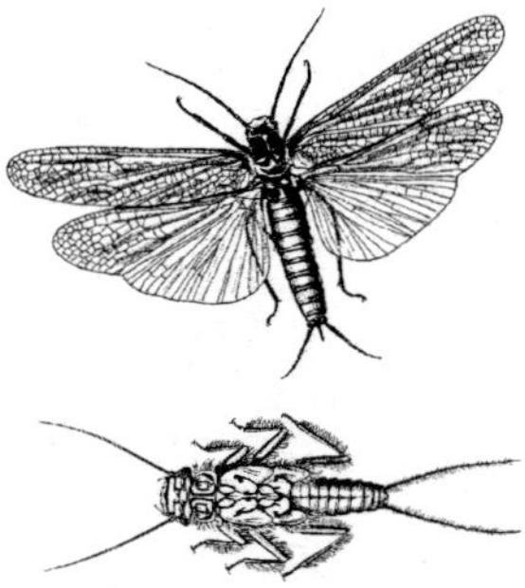

Stonefly and nymph

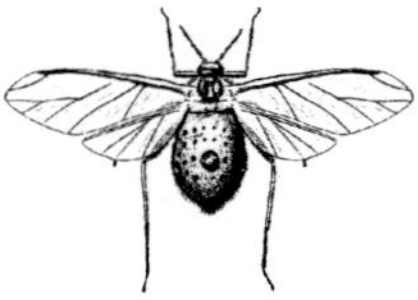

Aphid

Spittlebug

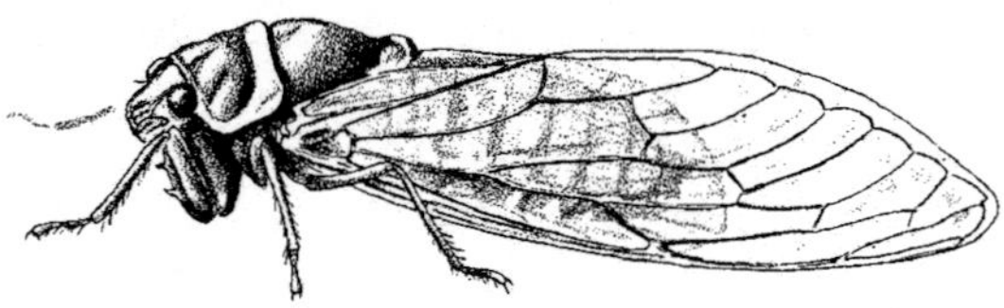

Cicada

tative of this order in the form of the ubiquitous introduced European earwig (Forficulidae: *Forficula auricularia*) occurs in the reserve. It can often be found inside the empty seed capsules of spice bush (*Calycanthus occidentalis*) and other moist dark places.

**Stoneflies** (Order Plecoptera). Stonefly nymphs (Perlodidae: *Kogotus alameda*) can be found in the bottom of the creek alongside mayfly nymphs. Although both groups look quite similar, they can be differentiated on the basis of tail number: while mayfly nymphs almost always have three, stonefly nymphs have two. The medium- to large-sized adults have two prominent sets of wings that fold like a tent above their abdomen, but retain their characteristic tails. Both nymphs and adult forms display a long, segmented abdomen.

**Aphids, leafhoppers, and cicadas** (Order Homoptera). The members of this group are incredibly diverse in form, from the soft-bodied, teardrop-shaped aphids with membranous wings to the slender, jumping leafhoppers to the froth-producing (juvenile) spittlebugs. Seek these small, delicate insects in trailside shrubs and other types of vegetation, and bring a hand lens for better viewing. Another member of this group is more often heard than seen. The electric buzz heard on hot summer afternoons is emitted by male cicadas (Cicadidae: *Platypedia* spp.), swift and elusive creatures that generally evade observation. There are several species on the reserve, the most colorful of which are black with red wing veins.

**True bugs** (Order Hemiptera). This group of insects is easy to recognize. The thickened, brightly colored forewings and membranous hindwings often result in a decorative X across the back of the bug. All true bugs have sucking mouthparts, and not surprisingly, many are predators. Many species are also aquatic. The most obvious riparian insect of the reserve is the water strider (Gerridae: *Gerris remigis*), which preys on unfortunate terrestrial insects that fall into the water. Water striders keep their stride on the water surface with their long middle and hind legs; however, look closely and you will also notice their short front legs, used to capture prey. Water boatmen, backswimmers, and giant water bugs spend most of their lives swimming in open water or clinging to submerged vegetation. Water boatmen (Corixidae: *Graptocorixa californica* and *Sigara mckinstryi*) have elongated oval bodies and oar-like hind legs. Backswimmers (Notonectidae: *Notonecta undulata*) look quite similar, but, as the name implies, they swim on their backs. Also, water boatmen eat algae and do not bite humans, whereas backswimmers are predatory and will give you a bite like a bee sting if you give them the opportunity. The giant water bugs of Cold Creek, amusingly named toe biters (Belostomatidae: *Abedus indentatus*), are indeed capable of inflicting a long-lasting bite. They look like large, broad water boatmen but do not have oar-like hind legs. Female toe biters lay their eggs on the backs of the males, so if you see an egg-clad bug by the creek, it is probably a giant water bug.

Terrestrial species have a variety of habits. Toad bugs (Gelastocoridae: *Gelastocoris* spp.) hop along the shore in search of insect prey. Stink bugs (Pentatomidae) also occur in the reserve, and, as their name suggests, they secrete a noxious odor if disturbed. Green stink bugs (*Chlorochroa* spp.) are herbivorous and cryptically colored to match the plants they feed on. The spotted assassin bug (Reduviidae: *Rhynocoris ventralis*), a medium-sized, eye-catching black and red or tan bug, is a sit-and-wait predator. It waits patiently for their prey to come to them, and then pounces on them.

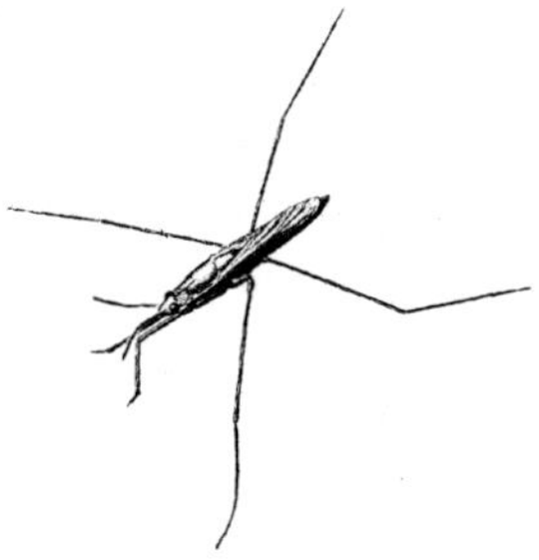

Waterstrider

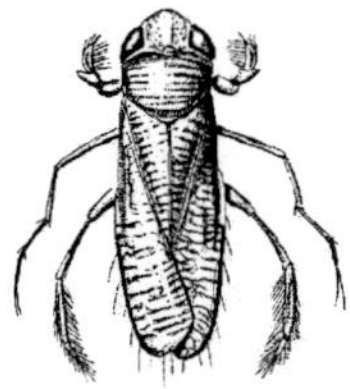

Water boatman

Backswimmers

Toad bug

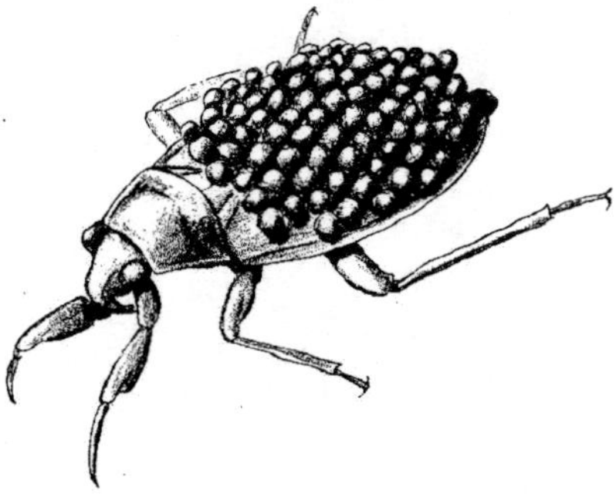

Giant water bug with eggs

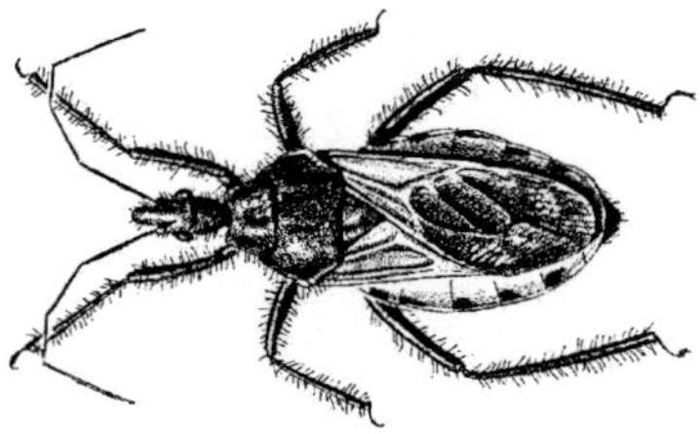

Spotted assassin bug

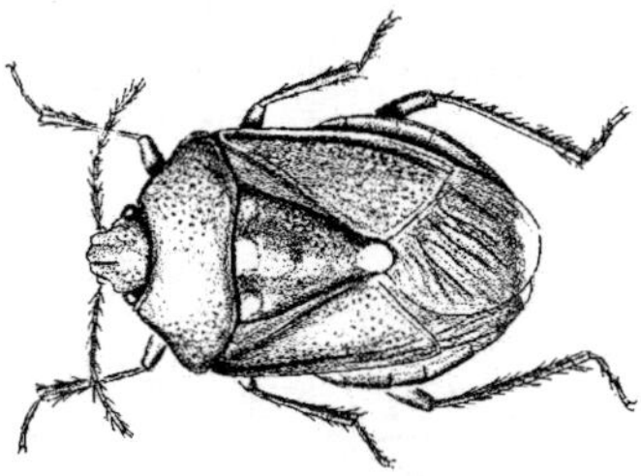

Stink bug

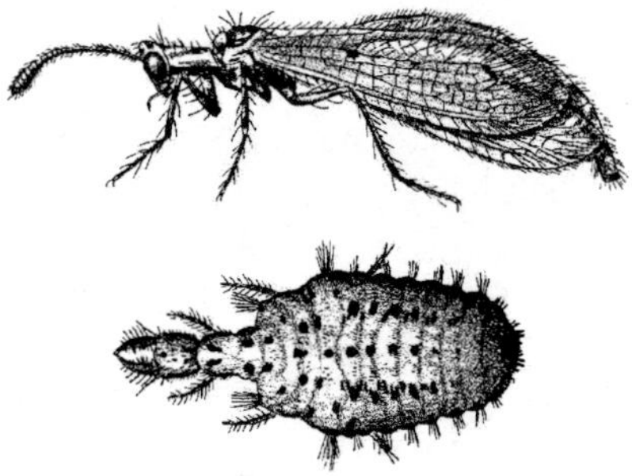

Antlion and larva

Lacewing

## INVERTEBRATES

**Lacewings and antlions** (Order Neuroptera). Adults of this group have two pairs of long, thin wings whose veins criss-cross in a quasi-checkerboard pattern not unlike a map of suburbia. The wings fold over their backs and often extend the length of their bodies. Green lacewings (Chrysopidae: *Chrysopa* spp.) are among the most common insects in the reserve. The golden-eyed adults flutter about on intricate, gossamer wings that belie their predaceous nature. Other related predators lurk at your feet. When walking on sandy or loose, dry soil, keep an eye out for conical pits. Larval ant lions (Myrmeliontidae) dig these pits and wait at the bottom for unfortunate insects to tumble in. Adult ant lions look superficially like gray damselflies with clubbed antennae.

**Beetles** (Order Coleoptera). The most diverse group of animals on the planet, the beetles can be found in almost every conceivable microhabitat in Cold Canyon, from aquatic to terrestrial. Look for their characteristic oval body shape and non-overlapping hard wing covers armoring their backs. In flight, beetles spread their wing covers, revealing clear, membranous wings underneath.

Several species of beetles are aquatic. Swarms of whirligig beetles (Gyrinidae: *Gyrinus plicifer*) can be found either on or below the surface of the quiet areas of the creek, usually twirling madly about en masse. The beetles have two pairs of compound eyes that enable them to watch for prey above and below the water simultaneously. To avoid collisions, the beetles rest their antennae on the water and can sense the presence of another object by the tilt on the water's surface.

Most terrestrial beetles spend their days hiding in the leaf litter or under rocks and come out only at night. One interesting ground beetle, the snaileater (Carabidae: *Scaphinotus* spp.), flaunts a long head and mouthparts used to eat its prey of snails and worms. Large, shiny black stink beetles (Tenebrionidae: *Eleodes* spp.) spend much of the day hiding under rocks and duff and have the amusing habit of walking about with their back ends raised much higher off the ground than their heads. Under harassment, stink beetles emit noxious fumes to deter predators.

Other beetles are diurnal and sport bright coloration. In autumn, spectacular swarms of ladybugs (Coccinellidae: *Hippodamia convergens*) converge in tree hollows and other protected sites. The only known endangered species inhabiting Stebbins Cold Canyon is the valley elderberry longhorn beetle (Cerambycidae: *Desmocercus californicus dimorphus*). The larvae of this species live in the core of elderberry trunks during their growth and maturation. Adult beetles emerge in the spring leaving a distinctive oval exit hole in the trunk that is about 1/4" in (1/2 cm) diameter. The adults live for only several weeks, are 1/2" to 1" in length, and have long antennae ('horns') that are as long as their bodies. The adult males and females look different; the males are mostly red with black spots, while females are mostly black with red around the edges.

**Butterflies and moths** (Order Lepidoptera). Butterflies and moths are among the most recognizable and beloved of all insects. They sport large, often showy wings, which are covered with tiny scales arranged like shingles on a roof. Most Lepidoptera sightings during the day at Cold Canyon are butterflies, though several intriguing day-flying moths occur on the reserve as well. In general, butterflies rest with their wings folded above their backs and have clubbed antennae, while moths fold their wings down along their bodies like lacewings and have feathery antennae; however, the exceptions are numerous! A spec-

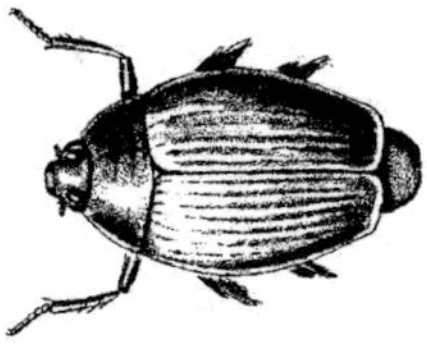
Whirligig beetle

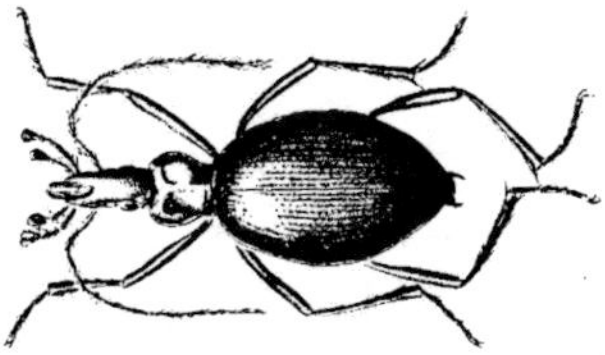
Snaileater

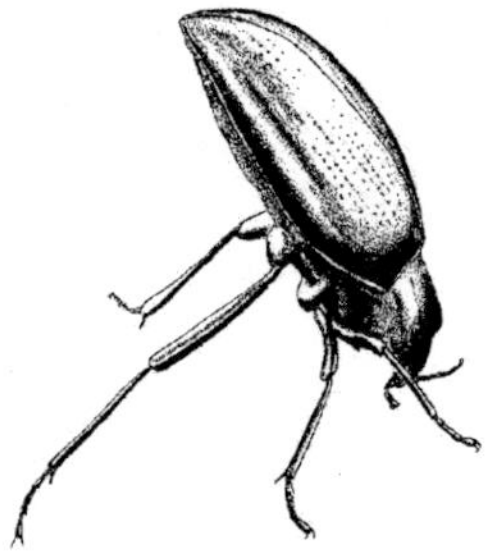
Armored stink beetle

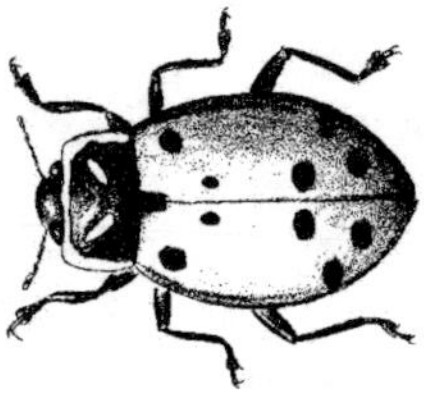
Ladybird beetle

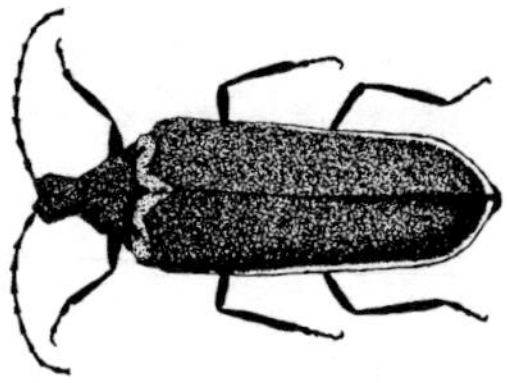
Valley elderberry longhorn beetle

Pipevine swallowtail

Western tiger swallowtail

California sister

Buckeye

tacular diversity of butterflies of all different shapes, colors, and sizes occurs in the reserve. Many are showy and can be spotted flying along the path, sipping nectar at flowers, or mudpuddling near the stream crossing.

**Pipevine swallowtail** (Papilionidae: *Battus philenor*). The pipevine swallowtail, a spectacular large, black butterfly with an iridescent green shimmer, courses up and down the canyon walls, especially in the spring. Its large black caterpillar, covered with red fleshy spines and spots, can be found eating California pipevine (*Aristolochia californica*) in the canyon bottom.

**Western tiger swallowtail** (Papilionidae: *Papilio rutulus*). The bold markings of the large and showy, yellow and black Western tiger swallowtail will catch even the disinterested eye. The larvae feed on willow, so the adults can often be found in the creek bottom.

**California sister** (Nymphalidae: *Adelpha bredowii*). Another large and conspicuous species, the California sister is brown with a diagonal white stripe and a large orange eye-catching blotch near the tip of each forewing. It can be spotted soaring lazily among the oaks.

**Buckeye** (Nymphalidae: *Junonia coenia*). In autumn, the buckeye, a medium-sized brownish butterfly with intricate, multicolored "eyespots," may be extremely numerous on blossoms of coyote brush.

**Chalcedon checkerspot** (Nymphalidae: *Euphydryas chalcedona*). Look for the medium-sized, strikingly patterned, black, white, and red chalcedon checkerspot basking in the spring sun.

**Common duskywing** (Hesperiidae: *Erynnis propertius)*. The common duskywing, a medium-sized, dark, common butterfly, is often initially mistaken for a moth because of its heavy body and habit of holding its front wings up like a butterfly's but its back wings down like a moth's.

**Sara orange-tip** (Pieridae: *Anthocharis sara*). Another lovely sight in early spring is the flash of color from the Sara orange-tip, a white butterfly whose wings are, as the name states, tipped with orange.

**Acmon blue** (Lycaenidae: *Plebejus acmon*). The acmon blue, a beautiful delicate blue butterfly the size of a child's thumbprint, often can be found in a virtual cloud around coyote brush in the same season as the buckeye.

**Echo blue** (Lycaenidae: *Celastrina echo*). A smaller but equally easy-to-spot butterfly is the echo blue, a shimmering blue butterfly that flits madly about in the spring and early summer.

**Moths** are also abundant in Cold Canyon; however, most of them are only flying about at night. One distinctive day-flying moth species, common in the reserve in the spring, is

Chalcedon checkerspot

Common duskywing

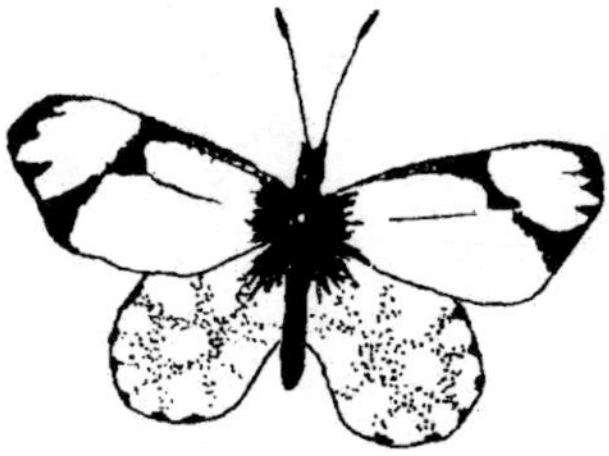

Sara orange-tip

Acmon blue

Echo blue

Bumblebee hawkmoth

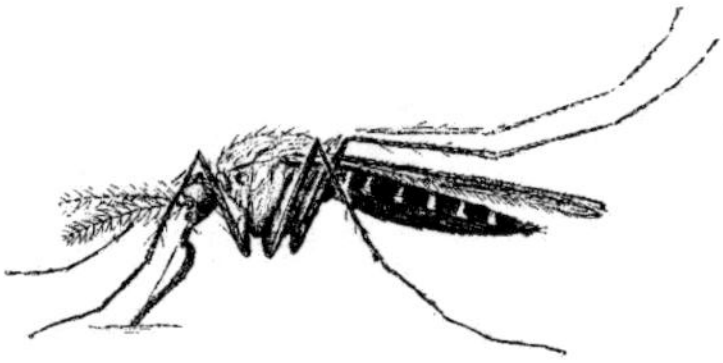

Mosquito

Black fly

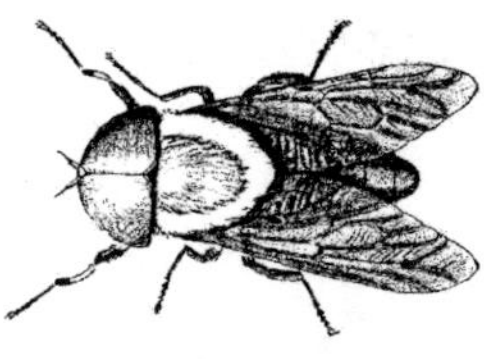

Horse fly

Snipe fly

the green bumblebee hawkmoth (Sphingidae: *Euproserpinus phaeton*), a convincing mimic of its namesake. In the late spring, grassy sunlit spots flicker with the presence of the three-striped longhorn (Incurvariidae: *Adela trigrapha*). The hair-like antennae of this tiny, silver-striped moth are several times its body length, giving it a distinctive appearance.

**Flies, gnats, and midges** (Order Diptera). If you cannot tell whether the insect buzzing across the path is a fly or a bee, carefully count its wings. If it has two sets of wings, you are probably looking at a bee or wasp; if it has only one set, it is a fly. Members of this order tend to be small, and their wings are semi-transparent.

Many flies are maligned because they bite unwary hikers. The most infamous blood-suckers are mosquitos (Culicidae: *Culex pipiens* and *Anopheles freeborni*), which develop as aquatic larvae in Cold Creek's pools and metamorphose into adults in late spring. Black fly larvae (Simuliidae: *Prosimulium* spp. and *Simulium* spp.) can also be found in abundance in Cold Creek. Club-shaped black fly larvae form dense black mats on many rocks in the creek. Although you will have to seek out the larvae in order to see them, the adults will be glad to come to you. Adult females must feed on blood to produce eggs, and they give a nasty bite. Adult female deer and horse flies (Tabanidae: *Tabanus* spp. and *Chrysops* spp.) do the same but can be differentiated from black flies because the Tabanid flies are brown and much larger. Many have brilliantly colored compound eyes and are known as "green-eyed monsters." Deer flies have black or brown patterns on the wings, while horse flies do not. Also among the biting flies of the reserve are the snipe flies (Rhagionidae). Common snipe flies (*Rhagio costatus*) attack exposed human limbs and inflict a painful wound. Adults are just under 1 inch (2 cm) in length, relatively hairless, and dull orange in color. Their natural hosts are probably the many non-human mammals of the reserve.

Fortunately, most species of flies are harmless to humans. Among the most primitive and unusual of the flies are the crane flies (Tipulidae), which are sometimes referred to as mosquito hawks even though they are not related to mosquitos and do not suck blood. However, like mosquitoes, they are nearly ubiquitous in cool, moist microhabitats. Common dance flies (Empididae: *Rhamphomyla* spp.), which have small heads and usually have smoky wings, can be found in the springtime. If you look at the males closely, you may discover the small, frothy prey balls they carry, which they present to the females as part of a courtship ritual. This dipteran equivalent of an engagement ring may contain a variety of smaller insects captured by the males.

Other flies are important pollinators. Many flower visitors that appear on first glance to be bees are in fact bee flies (Bombyliidae), which have evolved to look like bees, presumably to avoid hungry predators. These fascinating flies, delicate and fuzzy as adults, take flower nectar and pollen as nourishment. One of the most common species in the reserve is the greater bee fly (*Bombylius major*). They are bulbous, dark, fuzzy flies, and are often seen hovering over flowers or basking in moist, open areas near the trail. The front of each wing has a sharply margined dark brown blotch, and the rear of each wing is transparent.

**Sawflies, wasps, bees, and ants** (Order Hymenoptera). This order contains an exciting diversity of insects with a huge range of appearances and behaviors. Most hymenopterans have two pairs of wings, and a constricted abdomen at its attachment point with the thorax. Many species also have nasty stings. However, the most primitive groups are harmless.

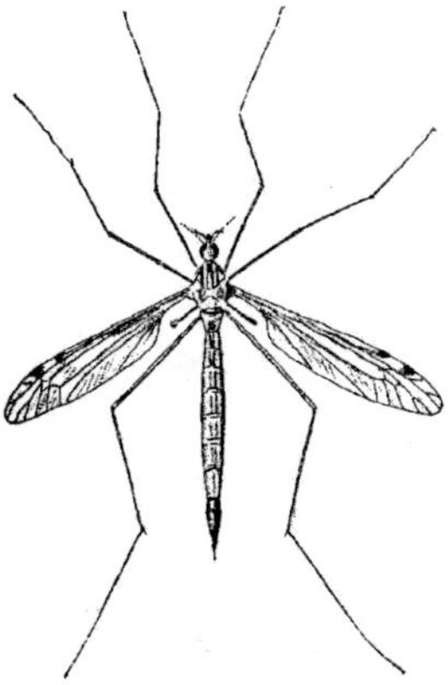

Crane fly

Dance fly

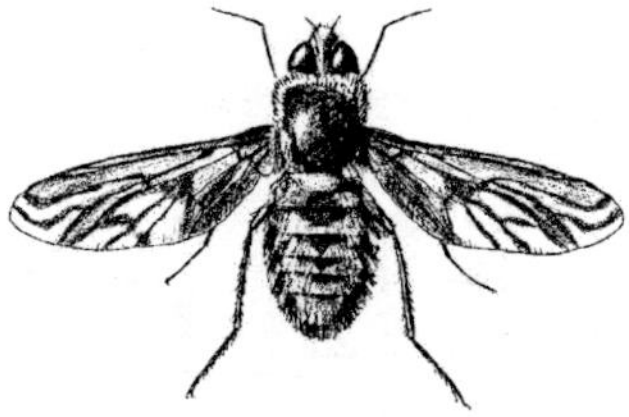

Bee fly

Sawfly

Oak gall wasp

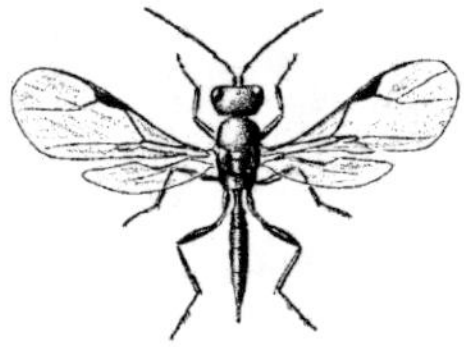

Aphid parasitoid wasp

Velvet ant

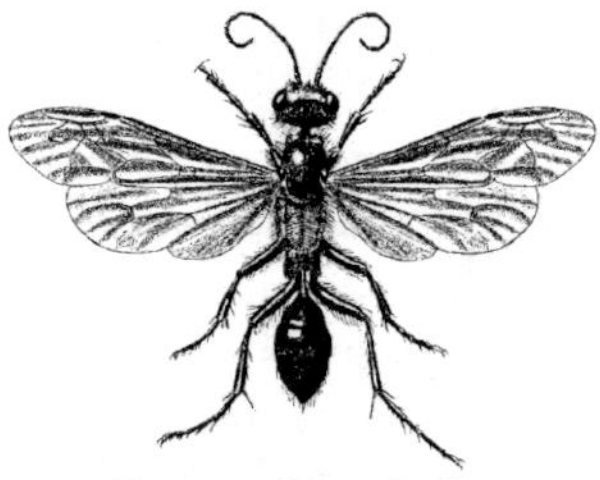

Green cricket hunter

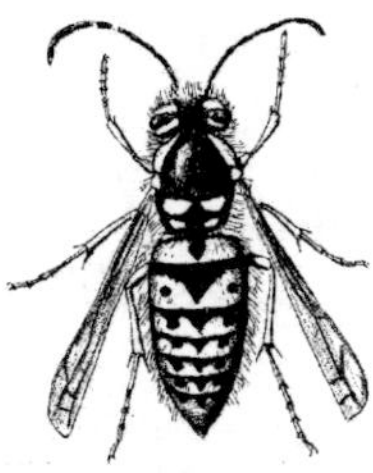

Yellowjacket

Paper wasp

Sawflies (Tenthredinidae: *Tenthredo* spp.), medium-sized, stout-bodied insects that attack plants, are present but not common in the reserve. Many species in this group exhibit parthenogenesis, wherein the females produce eggs without mating. All of the offspring are consequently female, and therefore many of these species have no males. The adults are rust-red to black with a large, rectangular head. The juveniles look caterpillar-like and feed on leaves, often gregariously. Another important group of hymenopterans in Cold Canyon and throughout California is the gall wasps (Cynipidae). The most reliable place to find gall wasps is on oak trees, where they occur in bewildering diversity. Most species lay their eggs in oaks, and chemicals released by the eggs or larvae induce the formation of tumor-like galls, whose shape is usually species specific. Depending on the wasp species creating the gall, it may protrude from the branches or the leaves of the oak. Remember that oaks reproduce by acorns, so any non-acorn protrusion you see on an oak, no matter how much it may look like an apple or some other fruit, is a gall.

The parasitoids, a diverse group of hymenopterans, feed on other insects without initially killing them outright. Parasitoids (Braconidae and Ichneumonidae) have interesting life histories, wherein the adult female deposits her eggs in the egg, juvenile, or adult form of another insect. The eggs hatch and the larvae then feed on the other insect, eventually killing it, but often leaving its skin intact. The carefully observant naturalist may see an adult parasitic wasp then emerge from a beetle larva, an adult aphid, or some other insect. Another common sign of parasitoid activity is rows of white parasitic wasp cocoons lining a caterpillar's back.

Many hymenopterans are wasps, predatory insects with a noticably pinched abdomen and a nasty stinger. One of the more visually appealing wasps is the velvet ant (Mutillidae), which looks like a furry ant. The females are wingless and scuttle along the ground; males look similar to the females but have wings. The "fur" on their backs is often brightly colored red, orange, or yellow. Velvet ants are solitary and parasitize other insects, especially ground-nesting wasps and bees. Another solitary group of wasps, the thread-waisted or digger wasps (Sphecidae), are primarily ground-nesting predators. The green cricket hunter (*Chlorion aerarium*), a metallic-green monster up to 1.5 inches (4 cm) long, can occasionally be seen dragging paralyzed crickets along the ground to its cloistered young.

Other groups of wasps are social, and form colonies or hives. Keep a lookout for the distinctive gray spherical nests of paper nest wasps (Vespidae). Paper nest wasps, the only paper-makers on the planet besides humans, mix wood pulp with saliva to create their nests. Two different black-and-yellow genera are common in the reserve. Yellowjackets (*Vespula* spp.) are aggressive, attacking nearly everything (including the meat in your sandwich), and deliver extremely painful stings. Paper wasps (*Polistes* spp.) are also predatory, but they are less aggressive and have slightly less troublesome stings. Yellowjackets often nest in abandoned rodent burrows and in hollow trees, while paper wasps hang their nests off of exposed tree limbs.

Like wasps, bees are often social and also have stingers, but bees are much less aggressive and are primarily pollinators. Two bumblebee species (Apidae: *Bombus edwardsii* and *B. vosnesenskii*), which are densely hairy yellow and black social bees, are extremely important members of the community because they are the sole pollinators for many native flowers. Without them, Cold Canyon's vivid spring wildflower display would be much less spectacular, and the food sources of certain insects and vertebrates would disappear. Another interesting and common group of bees are the metallic leafcutter bees (Megachilidae: *Osmia lingnaria*), which are bright metallic blue and live solitarily. Sociality is carried to an extreme in the introduced European honeybee (Apidae: *Apis mellifera*). Members of this species live as different castes: queens produce eggs, while sterile workers search for flowers, tend and defend the hive, and produce honey.

The most social of all hymenopterans are the ants (Formicidae). Ants can be pivotal members of natural communities, and the huge colonies of this group scavenge large quantities of organic matter including plant seeds and dead and living insects. One of the more common ants in Cold Canyon is *Messor andrei*, a seed-harvesting species with no common name, whose workers are black and medium sized. In the summertime these ants consume prodigious numbers of plant seeds and leave large accumulations of seed chaff around their nest entrances. Several species of carpenter ants (*Camponotus* spp.) also abound in the reserve. One common ground-dwelling species (*Camponotus semitestaceus*) nests both in rotten logs and directly in the soil. The workers of this species are orange-brown in color, with a contrasting darker head, and vary in size from medium to quite large (0.25-0.75 in. or 2-8 mm). The reserve is also notable for its lack of non-native ant species, such as the pestiferous Argentine ant (*Linepithema humile*), which is a serious threat to the continued diversity of native ants.

**Bumblebee**

**Leafcutter bee**

**Honeybee**

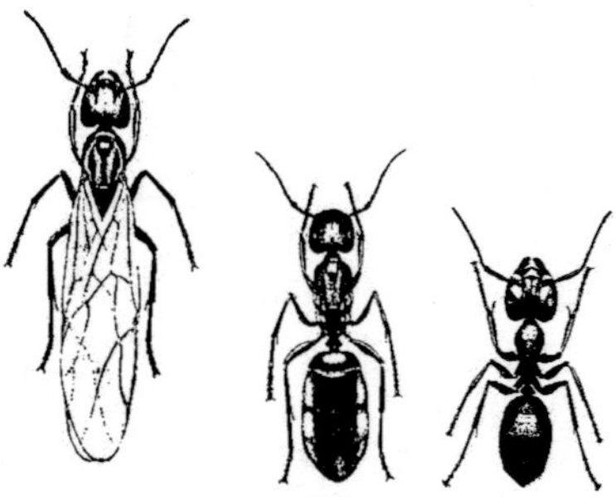

**Winged adult, wingless adult, and worker forms of the carpenter ant**

# Vertebrates of Cold Canyon

The sign at the entrance to Cold Canyon warns that you are about to enter mountain lion country, but despite claims of mountain lion encounters in the logbook, most of the animal action at the reserve is low to the ground where reptiles, mammals, and birds scurry into the shrubs at the sound of approaching footsteps. Frequently, the signs of animal activity, such as songs, tracks, and scats, are quite detectable while the animals themselves remain elusive. Observing wildlife in Cold Canyon requires patience and persistence.

The reserve's canyons and hillsides are, in fact, home to scores of vertebrates. At least 10 species of snakes, including the western rattlesnake, live in the canyons alongside salamanders, frogs, and lizards. In total, there are at least 25 species of reptiles and amphibians that make this their home, but with the possible exception of a few western fence lizards, most go unnoticed by the day hiker. So, too, do the 40-odd species of mammals that are active mostly only during the hours of darkness. While many of the canyon's resident birds are active by day, most stay hidden in the brush. Other birds are nocturnal and many others are not even permanent residents, using Cold Canyon as a rest stop during migration.

## Signs of Wildlife

While many vertebrates will elude the notice of even patient observers, most of these animals will leave telltale signs of their presence. The most common signs of mammals are their tracks. In the rainy months, the ground is soft enough to preserve the impressions of animals' feet. Because many mammals are elusive and nocturnal, their tracks are often the best way to determine whether they inhabit the area. Another common sign of mammal activity is the presence of scats, often found along the trail. Like tracks, scats can also be used to identify the presence of particular species. For example, fox scats are small and black and are often carefully placed by the fox in the middle of a rock as a territorial marker. Coyote scat can often be found in the center of the trail, and a close look may reveal the seasonally shifting diet of the animal – hair indicates rodent prey, and seeds and a purplish cast indicate a diet of berries.

**Black-tailed deer tracks.**

Other signs of animals include the houses they build. Most birds build nests, often hidden in the tangled branches of a chamise or manzanita bush, but sometimes on more exposed and observable branches. Dusky-footed woodrats build a complex house out of sticks and other debris. These impressive conical structures are up to 6 feet (1.8 m) tall and 8 feet (2.4 m) in diameter, with thick roofs and several passageways leading to multiple chambers.

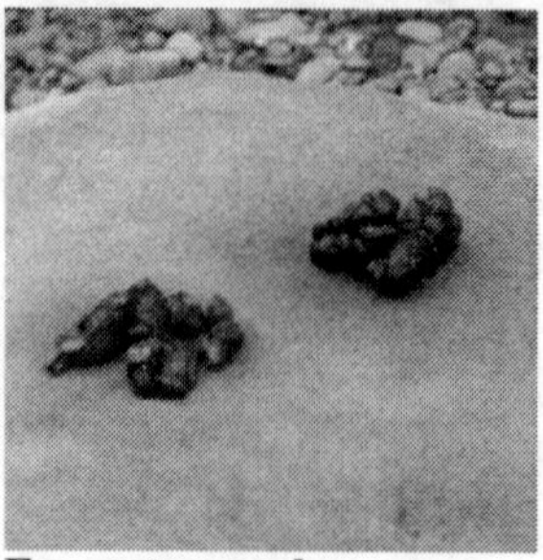

**Fox scat on rock.**

## Observing Wildlife

Here are some suggestions if you would like to observe the vertebrates of Cold Canyon. First, remember that many animals have much keener vision, hearing, and sense of smell than do humans. If you try to approach something moving in the brush, chances are that the animal will notice you and retreat, or freeze in fear and remain hidden in the bush. Often the best way to observe these animals is to sit quietly and wait. If you escape detection by the animal for several minutes, it may come out from hiding. Better yet, start your observations without the animals realizing you are there. Pick a quiet comfortable place in which you are relatively hidden. Because many animals can detect you simply by your smell, your observation place should be downwind of the area you want to observe. Use binoculars to observe animals from a greater distance than they can detect you.

The animals you observe will greatly depend upon the time of day you visit Cold Canyon. To the midday visitor, first impressions can be misleading. Nothing seems to be out and about. Except for the chatter of a few birds, a fleeting glimpse of a squirrel, or the skitter of a lizard, there simply does not seem to be any life in the area.

Come back at first light of morning, and you'll experience a completely different sensation. Tread your way up the canyon, away from the highway, and you'll hear what birders call the dawn chorus. You'll be surrounded by dozens of songs, coming from just as many bird species. During this time, it is much easier to spot particular birds because they often sing boldly from conspicuous perches and may even be attracted to you if you whistle softly.

Or arrive in the late afternoon and wait for dusk to creep up the canyon slopes. At sundown, the coyotes begin yipping up canyon as owls call from their tree limb perches. Harvest mice and meadow mice roam the grassy hills, searching for seeds and insects, and dusky-footed woodrats begin to stir from their houses, in search of food and more plant material to add to their houses. In addition, many amphibians and mammals (and at least one reptile, the night snake) are nocturnal and, as a result, go about their activities unseen by humans.

The season you visit Cold Canyon is as important as the time of day for observing particular species of vertebrates. Many more bird species are active during the spring and summer than in other seasons, and your chances of seeing rare birds are greatest during the early spring and fall migrations. Reptiles depend on the warmth of sunshine, so most are active in late spring, summer, and early fall. However, most amphibians require moisture to prevent dessication and, as a result, their activity peaks during the rainy season of late fall, winter, and early spring. Mammals are present year-round, but many are inactive during the winter.

## Fish, Amphibians, and Reptiles

Cold Canyon is home to a fair diversity of the "lower vertebrates" – especially amphibians and reptiles – but the secretive nature of these creatures makes it difficult for naturalists and hikers to appreciate them. There are two other reasons it might be difficult to observe the reserve's resident amphibians and reptiles. First, as previously described, certain species may be active only during particular times of the day or during certain seasons. Second, some amphibians and reptiles live in restricted habitats of Cold Canyon and are therefore difficult to locate. For example, whiptail lizards prefer the dense growth of chaparral and thus do not come into contact with most humans. Other reptiles are primarily fossorial (underground dwellers) and so rarely see the light of day. As a result of these factors, some species are suspected to occur in Cold Canyon but have not been confirmed with a positive identification.

Below are descriptions of the most common fish, amphibians, and reptiles found in Cold Canyon. A complete list of known and suspected species is located in the appendices (p. 115).

## *Fish*

**Rainbow trout** (*Oncorhynchus mykiss*). This native species is the only fish that might on occasion be found in the creek. During rainy years, Cold Creek swells in its banks and some rainbow trout are capable of moving up from Putah Creek. When the creek dries in the summer, permanent pools in Wild Horse Canyon can provide a home for these trout. In such situations, these fish may survive to reproduce – starting a new generation in Cold Creek. However, this is a rare occurrence. In most years, the pools dry up and trout are excluded from the area, so rainbow trout cannot truly be considered permanent residents of Cold Canyon.

## *Amphibians*

**California newt** (*Taricha torosa*). The California newt is a medium-sized salamander (5-7 in., or 13-18 cm, from head to tail), with textured skin that is tan colored on its back and orange on its belly. Newts have a complex and intriguing life cycle. In the pools of the creek, they hatch from eggs as tadpole-like juveniles. As the pools dry up, they metamorphose into their adult form and head for moist logs and holes in the ground. After the fall rains replenish the water in the stream, the adult newts return to the pools, metamorphose into an aquatic form, and reproduce. The best places to spot California newts are near the stream and on moist portions of the trail. In winter you can find the newts heading for the water, and in the spring you can find them heading back into the hills. At all times, newts move slowly and seem oblivious to other animals. Their nonchalance is likely due to their toxic skin – each gland contains noxious poison. They advertise this weapon with their bright orange belly, a warning recognized by most predators.

**California slender salamander** (*Batrachoseps attenuatus*). While these tiny (1-3 in., or 2-8 cm, long), brown, wormlike salamanders are most active during the winter and early spring, they can be found in moist places under rocks and logs as late as June. Unlike newts, slender salamanders do not spend any of their life cycle in the water. In the winter, they lay their eggs underground. These hatch in the early spring, and the young are born looking like smaller versions of their parents.

**Western toad** (*Bufo boreas*). Western toads are 2 to 5 inches (5-13 cm) long, brown, and covered with warty bumps. These bumps hold toxins that can easily kill a dog or other predator unlucky enough to bite a western toad. Toads spend their adult lives hiding in leaf litter and burrows and are usually active only at night. During the winter wet season starting in January, they come out of hiding to breed in Cold Canyon's pools. Males station themselves around the pools and attract females with weak, chirpy calls. Females lay fertilized eggs in the creek, and tadpoles hatch within several weeks. The tadpoles develop rapidly, metamorphose into tiny toadlets, and are out of the creek pools and hiding in the wooded slopes of the canyon by May.

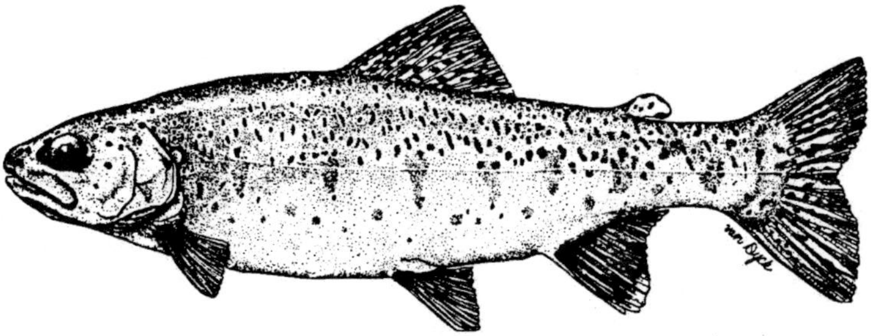

**Rainbow trout**

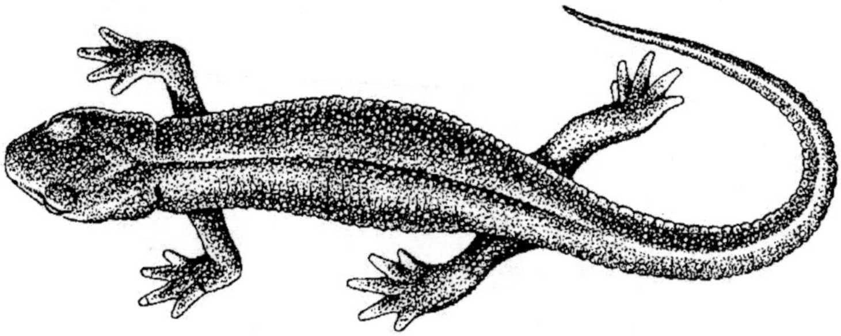

**California newt**

**California slender salamander**

**Western toad**

**Pacific treefrog**

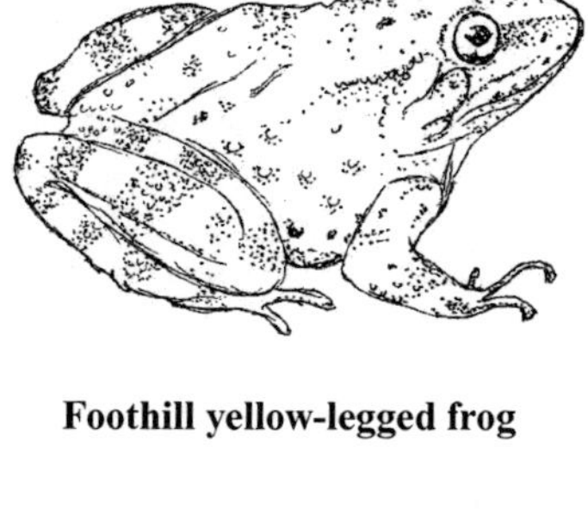

**Foothill yellow-legged frog**

## VERTEBRATES

**Pacific treefrog** (*Hyla regilla*). These small (1-2 in. or 2-5 cm) frogs are the most common frogs in California. They come in many colors, from green to tan to brown, and individuals can actually change color fairly rapidly. Fortunately, a dark eyestripe that does not change color distinguishes these frogs from all others in Cold Canyon. Contrary to their name, Pacific treefrogs generally stay near the ground, although their padded toes enable them to climb if they need to. Like western toads, Pacific treefrogs breed in the pools of Cold Creek from January to May, and, during this time, males attract females by calling. Their repeated *krek-ek* calls, which seem much too loud for such a small creature, are the most common type of frog call that Californians hear. Eggs are laid in streams and hatch into tiny black tadpoles, which rapidly metamorphose into the adult form.

**Foothill yellow-legged frog** (*Rana boylei*). These medium-sized (1-3 in. or 2-8 cm) frogs can be identified by their mottled back and yellow underside and their tendency to dive into pools and hide in the muck at the bottom when disturbed. Yellow-legged frogs breed from March to May and will stay in the vicinity of the creek the entire year. As a result, they are the first to return after the fall rains begin, and male yellow-legged frogs initiate the breeding season with their grating, guttural calls. Of all the amphibians in Cold Canyon, the foothill yellow-legged frog is perhaps the most endangered. Although they are locally abundant, yellow-legged frogs have suffered huge declines across the state, largely due to habitat destruction and introductions of non-native predatory fish.

### *Reptiles*

**Western fence lizard** (*Sceloporus occidentalis*). These "blue-bellies" are the most common reptiles in Cold Canyon and probably all of California. They are 4 to 6 inches (10-15 cm) long from head to tail, with brown and black blotches or lines on their spiny-scaled back. They can be seen along much of the trail, and they colonize the creek bed when it dries up. Most lizards, especially the males, sport blue patches on their throats and the sides of their bellies. They use these to communicate via "push-up" displays, during which they lift the front of their bodies repeatedly with their front legs.

Western fence lizards provide a surprising service for humans and potentially other mammals in the canyon. Deer ticks, which carry Lyme disease, bite lizards as well as humans. But unlike humans or any other vertebrate, western fence lizards produce a substance that kills Lyme disease. When a tick attaches to a lizard, the substance in the lizard's blood enters the tick and kills the Lyme disease bacteria inside the tick. As a result, areas that have lots of western fence lizards have a low prevalence of ticks infected with Lyme disease. Therefore, a large population of lizards reduces the possibility of Lyme disease infection in humans.

**Western skink** (*Eumeces skiltonianus*). Western skinks are small (3-7 in. or 8-18 cm), smooth-scaled lizards that undergo a change of color as they grow into adults. Young skinks are brightly striped and have showy, almost neon blue tails, while adults are generally brownish with an orange head and tail. Herpetologists speculate that the blue tails of young skinks help them escape predators. If the predator is quick enough to catch a young skink, it is attracted to the coloration of the tail. When grabbed, the tail readily breaks off and thrashes about with a life of its own, distracting the predator, while the skink quickly runs for cover.

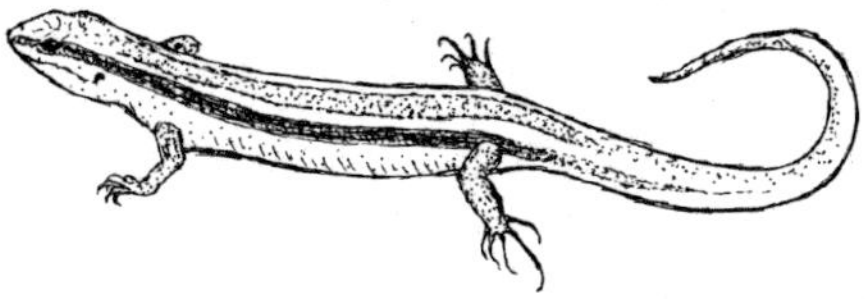

**Western skink**

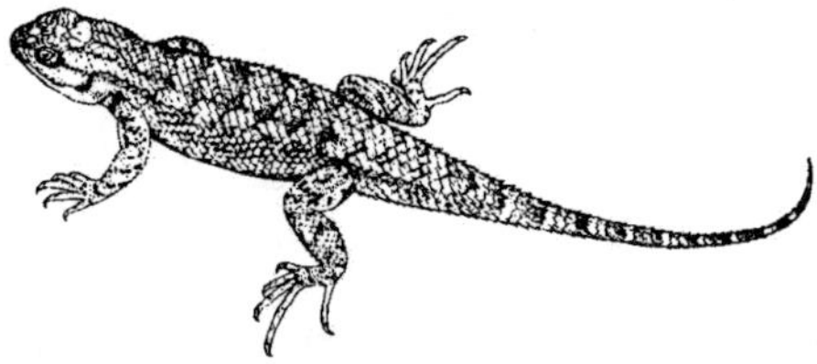

**Western fence lizard**

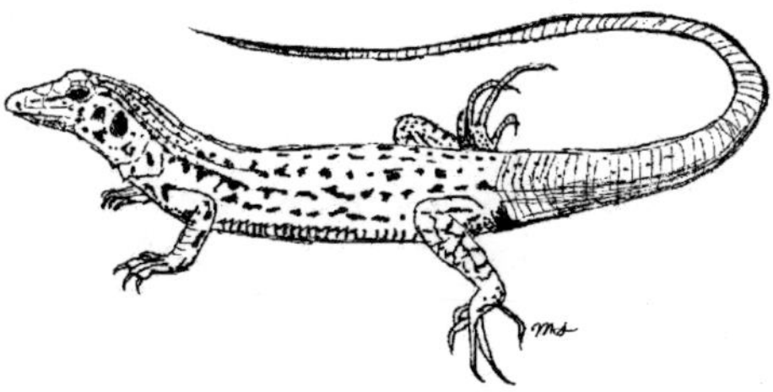

**California whiptail**

**Southern alligator lizard**

**Western yellow-bellied racer**

**Gopher snake**

**Western aquatic garter snake**

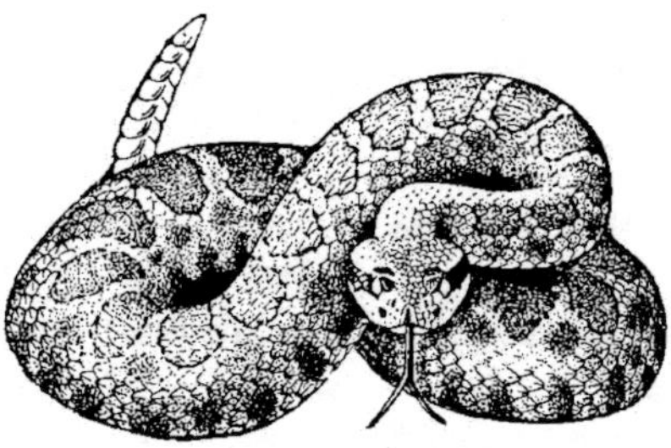

**Western rattlesnake**

## VERTEBRATES

**California whiptail** (*Cnemidophorus tigris*). Whiptails are the sprinters of the reptile world. These medium-sized lizards (4-8 in. or 10-20 cm), characterized by stripes and bars coloring their small scales, can run up to 30 miles an hour if pressed. Usually they are seen moving nervously underneath shrubs. Approach them, and they will rapidly skitter into the chaparral. Like skinks, whiptails are born with blue tails, which they lose as adults.

**Southern alligator lizard** (*Elagaria multicarinatus*). These are the largest lizards in Cold Canyon, ranging from 6 to 12 inches (15-30 cm) in length from head to tail-tip. Alligator lizards have smooth tan scales with darker bars. They are often found in grasslands chasing insects, small lizards, and young mice, but will also climb trees and eat bird eggs. They are aggressive lizards that will take a bite at your finger if cornered.

**Western yellow-bellied racer** (*Coluber constrictor*). These olive or brown smooth-scaled snakes can get up to 7 feet (2.1 m) long. They are well named, for they move at surprising speeds when chased. Racers are often found in grasslands and savanna. They prey on many small vertebrates, especially lizards.

**Gopher snake** (*Pituophis melanoleucus*). These snakes are 3 to 8 feet long (0.9-2.4 m) and have cream-colored, keeled scales with brownish blotches or stripes. They often look like rattlesnakes, and in fact they are known to imitate them. A defensive gopher snake will often rear up, hiss, and shake its tail in the leaf litter to simulate a rattle. While they have a nasty temper, gopher snakes are not venomous and kill their primarily mammalian prey by constriction.

**Common garter snake** (*Thamnophis sirtalis*). These thin, striped snakes range from 18" (46 cm) to almost 6 feet (1.4 m) in length. Most Common garter snakes at Stebbins have black scales, yellow stripes, and bright red skin between their scales. They are often found in Cold Creek's pools or basking on nearby rocks. They are excellent swimmers and will retreat to the water when alarmed. Garter snakes do not lay eggs, but instead allow the eggs to develop in their body and bear the young live.

**Western rattlesnake** (*Crotalus viridis*). This is the only venomous snake living in Cold Canyon and Northern California. It grows up to 6 feet (1.8 m) in length and is covered in gray or brown keeled scales. This snake uses its conspicuous rattle at the end of its tail to create a buzz-like warning noise when it is startled. It is active from April to October, but is most frequently encountered from May to July. Rattlesnakes are common in the reserve, but are most often seen on the dry hillsides rather than in the canyon bottoms. Although potentially dangerous, this rattlesnake is not aggressive and usually retreats unless cornered or provoked. Young rattlers tend to be more dangerous than adults, because young do not retract their fangs when they strike, and therefore are much more likely to inject large quantities of venom into the bite. Rattlesnakes eat all types of small mammals and lizards, but they are best adapted to capture small mammals and birds. Small pits located on their faces enable the snakes to detect temperature differences. Therefore, they are readily attracted to warm-blooded animals. Not surprisingly, they often hunt at dusk, when they can most easily detect these temperature differentials.

## Birds

The birds of the Cold Creek and Wild Horse drainages can be categorized into at least 34 families, 82 genera, and 114 species, one half (56) of which are known or suspected to nest in these two basins. By comparison, the terrestrial bird count for the entire state of California is about 235 regularly occurring species. Thus, the Cold Canyon basin contains nearly 50% of the total for the state, a most impressive number for such a small area. About one half of the recorded species either pass through the reserve during migration (transients) or migrate to the reserve to spend the winter (winter visitors). The descriptions below highlight the most common and more permanent residents. Use the species list in the appendices on pages 117 to 121 as a complete checklist. Birds are the only taxonomic group assigned official common names, so their names are capitalized, unlike the common names of other animals and plants.

**Turkey Vulture** (*Cathartes aura*). The Turkey Vulture teeters from side to side as it soars on long wings held up in shallow V. It uses both vision and smell, not in search of living prey, but in search of carrion, or carcasses of previously killed animals. Its plumage is black overall, though the undersides of its flight feathers have a silvery cast, and the red skin of its head (gray in juveniles) is unfeathered. Turkey Vultures lay their eggs in sheltered ledges on cliffs such as those found to the west above Cold Canyon.

**Red-tailed Hawk** (*Buteo jamaicensis*). Scanning the grasslands and savanna for a meal, large birds of prey such as the Red-tailed Hawk soar above the reserve on the updrafts that rise off steep canyon walls. Though they vary widely in coloration, most Red-tailed Hawks are a shade of brown above, with a lighter shade underneath. Darker markings fleck their broad wings and form a band across their belly, and adults exhibit their namesake rust-colored tail. The Red-tailed Hawk's piercing call, a familiar sound effect from movies and television, descends in pitch and is often heard in the spring as mated pairs perform acrobatic aerial courtship displays.

**California Quail** (*Callipepla californica*). A prominent, curly black head feather marks the shy California Quail. Quail can be found running about among shrubs that neighbor grasslands. They often travel in large coveys – groups usually composed of several adults and numerous young – in search of acorns, seeds, and fresh grass. Catching them by surprise often causes a loud burst of short flight, followed by a confused chorus of peeps as the birds run about to regroup. These monogamous birds produce large clutches of eggs, and the young that hatch are immediately able to feed themselves and run on the ground.

**Great Horned Owl** (*Bubo virginianus*). A voracious and silent hunter of the night, the Great Horned Owl will prey upon almost anything it can catch, including mice, woodrats, and even animals as large as skunks and pheasants. Most of the day, Great Horned Owls remain inactive, moving about only when they are discovered by humans or by an unruly mob of birds. However, during the early spring, males often hunt in the late afternoon to accommodate their hungry babies. As you hike near the oak woodlands, listen for the shrill alarm calls of jays and flickers; they might be advertising the location of a Great Horned Owl.

## VERTEBRATES

**Anna's Hummingbird** (*Calypte anna*). In metabolic overdrive, tiny hummingbirds hover by rapid wingbeats as they drink sugary nectar from flowers with their needle-thin bills. Exceptional in its ability to overwinter, Anna's Hummingbird is an abundant year-round resident at the reserve, while other hummingbirds are only present during their breeding or migration seasons. Both sexes of Anna's Hummingbird have metallic green upperparts and pale gray undersides with a green wash. When light is reflected from a certain angle, the male's head and throat feathers shine pink, and the female's throat is sparsely flecked with these iridescent feathers. The Anna's male sings a squeaky series of high chirps from a perch.

**Northern Flicker** (*Colaptes auratus*). The Northern Flicker is a loud resident of the canyon, chiding hikers with a *flick-a flick-a flick-a* call as it flies over the trail. Look for its white rump, its black-spotted belly, and the reddish undersides of its wings and tail. Flickers tend to fly with several rapid wingbeats followed by a glide. These birds nest in cavities they hollow out of dead trees with their stout beaks. These cavities are also often important for other bird species, including swallows, bluebirds, and screech owls.

**Nuttall's Woodpecker** (*Picoides nuttallii*). The series of rattling calls given by a Nuttall's Woodpecker often announces its presence high on the side of a tree. Its back is barred horizontally with black and white, and its head is black with white marks. The male also sports a red crown. Nuttall's Woodpecker uses its specialized bill to flake away bark as it forages for insects, to make a territorial drumming on tree trunks, and to excavate a nest cavity in a trunk or branch.

**Black Phoebe** (*Sayornis nigricans*). Closely associated with riparian habitat, the diminutive Black Phoebe can be seen perched on branches hanging over Cold Creek. From its perch, the Black Phoebe nervously flicks its tail and makes occasional flittery flycatching forays into the air. A solid black head, upperparts, and breast with a contrasting white belly give this bird a trim coloration. Two rising notes followed by two descending notes constitute its song.

**Western Scrub-Jay** (*Aphelocoma coerulescens*). Well known in California due to its abundance and brash behavior, the Western Scrub-Jay can scarcely be missed in the reserve. Its loud, grating call is given as a singular statement or in a series. The sky blue of its head, wings, and long tail surrounds its gray back and extends as a broken necklace into its white underparts. In autumn the Western Scrub-Jay switches from a diet of mostly insects and begins to collect acorns and cache them in the ground. By burying more acorns than it consumes over the winter, the Scrub-Jay effectively plants the next generation of oaks. It is believed that this relationship allows the oaks to disperse uphill, counteracting the tendency of acorns to roll downhill.

**Cliff Swallow** (*Hirundo pyrrhonota*). Flying and gliding about the steep canyon cliffs, Cliff Swallows seem to be forever in flight, rarely pausing to rest from their persistent insect foraging. These birds build nests out of daubs of mud on cliff faces. Because these birds are colonial, there are usually hundreds of nests placed right next to each other. Three

Turkey Vulture

Red-tailed Hawk

Great Horned Owl

California Quail

Anna's Hummingbird

Northern Flicker

Nuttall's Woodpecker

other swallow species are regularly seen in Cold Canyon, but only the Cliff Swallow sports a small buff-colored patch on its forehead.

**Oak Titmouse** (*Baeolophus inornatus*). The Oak Titmouse is a small, uniformly gray bird, with a pointed crest on its head. These birds flit among trees and shrubs, energetically gleaning insects and berries. Although shy, they can often be heard repeatedly whistling *tee-wit tee-wit*, the most common of their broad repertoire of songs and calls.

**Bushtit** (*Psaltriparus minimus*). Tiny but kinetic, the Bushtit is pale gray with a light brown crown and relatively long tail. Frantically moving through the chaparral and woodlands as they forage in winter, flocks of Bushtits keep in constant communication by a twittering of whistled chirps. In spring these contact calls continue between mated pairs as they build elaborate gourd-shaped hanging nests of plants and lichen woven together with silk from the webs of spiders and cocoons of insects.

**Bewick's Wren** (*Thyromanes bewickii*). Although several types of wrens live in Cold Canyon, the Bewick's Wren is the most conspicuous. These wrens are can be spotted flitting into the foliage where they hunt for insects and hide from predators, but they can often be coaxed out of the brush by whistling and *pssshing*. Bewick's Wrens have long tails with white edges, which they stick up in the air, and a white eyestripe extending across their face from their curved bill. When out of sight they will often sing a three-note *swee swee cheeeee* song. The first two notes are high pitched, while the third is a lower, longer trill.

**Wrentit** (*Chamaea fasciata*). The Wrentit is one of the many small birds with drab coloration that blends in with the brushy environments of the reserve. On the rare occasion when it emerges from the shelter of dense undergrowth, one can see the Wrentit's dark gray-brown plumage, white irises, and long tail, which is often held up at an angle. The Wrentit is more often heard than seen, giving its characteristic "bouncing ball" song, a descending series of notes that becomes more rapid until running together as a trill.

**Yellow-rumped Warbler** (*Dendroica coronata*). While most North American wood warblers migrate to the tropics, the hearty Yellow-rumped Warbler can be found here in abundance as a winter visitor. Gathering in flocks, wintering Yellow-rumped Warblers give a sharp chirp as a contact call. Though variable by race, sex, and season, plumage follows a basic pattern of gray upperparts with patches of yellow above the tail, on the sides under the wings, and on the crown. The "Audubon's Warbler" form, more common in the West, has white crescents above and below the eye and, in males, a yellow throat. The "Myrtle Warbler" form can also occasionally be spotted in the canyon and can be distinguished by a white eyebrow stripe and white throat extending to the sides of the neck.

**Spotted Towhee** (*Pipilo maculatus*). The Spotted Towhee prefers chaparral or the understory of woodlands and is often heard scratching with both feet simultaneously as it combs the leaf litter for seeds and insects. Black on the male – brown on the female – covers the upperparts and forms a hood over its head and upper breast, accented by red eyes, white wing bars, and white spotting above the wings. This appearance long kept the Spotted Towhee classified with its eastern relative, the Rufous Towhee, but recently it has been

Black Phoebe

Western Scrub-Jay

Cliff Swallow

Oak Titmouse

Bushtit

Bewick's Wren

recognized as a separate species because evidence suggests that the two populations do not interbreed. A buzzy trill is the song given by the male, usually from a prominent perch. The call of the species is a scratchy, rising slur.

**White-crowned Sparrow** (*Zonotrichia leucophrys*). When winter brings harsh weather and a scarcity of available seeds to their breeding territories high in the mountains, populations of White-crowned Sparrows move into suitable habitat in the Central Valley and interior Coast Range, including the woodlands of the Stebbins Cold Canyon Reserve. Identify this small ground-feeding sparrow by its black-and-white striped crown (brown striped on immatures) and pink to orange bill. The White-crowned Sparrow's song consists of one to three clear whistles followed by a trill, and they vocalize all winter. A close listen distinguishes the dialect sung by wintering White-crowned Sparrows at Cold Canyon from that of the resident populations on the California coast.

**Lesser Goldfinch** (*Carduelis psaltria*). A flash of yellow in the trees above the trail may mark the flight of a Lesser Goldfinch. Male Lesser Goldfinches sport bright yellow breasts and black capes that cover their heads and run down their backs. Females are olive-colored, with dark wings and tails. These birds search for insects among the live oak and riparian woodlands, singing a paired *tee-yee* call to communicate with each other.

**House Finch** (*Carpodacus mexicanus*). The House Finch is a common resident of Cold Canyon. These birds can often be seen searching for seeds and fruits, which make up most of their diet. A reddish breast, forehead, and rump readily distinguish the males. Females are gray-brown above and streaked below. House Finches brighten the afternoon canyon quiet with an extended, lilting musical song. In the late 1800's, these birds were prized by pet sellers in the eastern United States. After legislation outlawing collection of these birds was passed, hundreds of finches were released by pet store owners, and some of these birds became established in New York. This population spread across the east, outcompeting the native Purple Finch (*Carpodacus purpureus*). Now House Finches can be found almost everywhere east of the Mississippi, in addition to their native, semi-arid habitat in the West.

## Mammals

Most visitors to the Stebbins Cold Canyon Reserve may give little thought to the rich mammalian fauna that lives in the region. After all, most mammals are nocturnal and those that are active during the daylight hours are usually quite secretive, rarely showing themselves to human visitors. However, it would be a serious mistake to dismiss the mammals just because they are less conspicuous than their avian neighbors. Ranging in size from the quarter-ounce shrew to the large, lumbering black bear, over 40 species of mammal call this area home for much or all of their lives. Many factors contribute to this diversity, but particularly relevant is the location of the reserve at the intersection of the valley habitat to the east and the coastal mountain habitat to the west.

It is, therefore, the diversity of habitats – whether open, hot and dry, or shady, cool and dense – that dictates to a large extent the composition of mammal species at any specific site. On the grassy hills to the west, harvest mice (*Reithrodontomys megalotis*) and Botta's pocket gophers (*Thomomys bottae*) are the most common mammalian residents. Drop

Wrentit

Yellow-rumped Warbler

Spotted Towhee

Lesser Goldfinch

White-crowned Sparrow

House Finch

down into a riparian canyon in the middle of this grassland where the vegetation changes to a more shade-dominated mix of plant species and where the soil layer gives way to rocks and gravel, and you will find an abundance of two species of field mice (*Peromyscus maniculatus* and *P. boylii*) and very few harvest mice and pocket gophers. Walk over to the west side of the canyon, where shade dominates the landscape, and you will not only find these same two species of *Peromyscus* in high numbers, but also the dusky-footed woodrat (*Neotoma fuscipes*), whose presence is revealed by the large, stick houses that it builds. A number of bat species find the canyon attractive as well. Some roost in the thick shrubs and tree branches, while others find safe daytime lodging in the cracks and caves of the two rocky ridgelines that traverse the reserve.

A vigorous assemblage of small mammal species means that a number of predators can also be found in the canyon. Coyotes (*Canis latrans*) and gray fox (*Urocyon cinereoargenteus*) are two of the most abundant predators, and they make their presence best known by the piles of scat that they deposit on the many animal trails that crisscross the reserve. It is from these scats that field biologists can quickly determine the array of prey in the area, because bones and teeth, diagnostic in species identification, remain in the scat.

The reserve's three largest mammals are black bear (*Ursus americanus*), mountain lion (*Puma concolor*), and mule deer (*Odocoileus hemionus*). Both bears and mountain lions are most often recorded by the tracks that they leave in the mud along either Cold Creek or Wild Horse Creek. While they are certainly not numerous, usually at least one or two sightings are reported each year. Deer can be found throughout the canyon, as the rich and diverse array of tasty and nutritious plants offer year-round browse.

The most common mammals in the reserve are described below. A complete list of all mammals that have been observed in Cold Canyon can be found in the appendices (p. 122).

**Virginia opossum** (*Didelphis virginiana*). The opossum is the only marsupial in North America. Like all marsupials, it raises its young in an abdominal pouch. This pouch may be the key to its abundance – an opossum can simultaneously have a batch of young developing in the pouch and an additional set on its back, and thereby produce a huge number of young in its lifetime. Another key to its abundance is its omnivorous habits. The opossum eats fruits, insects, carcasses, and garbage, and these scavenging habits have made it a common city resident. Observing an opossum is difficult, as they are active only at night. If you are lucky enough to stumble on one, you may witness its classic ability to play dead. In fact, the opossum is not playing – its immobility is caused by a fright reaction.

**Raccoon** (*Procyon lotor*). A characteristic dark mask and banded tail distinguish the ubiquitous raccoon. Raccoons are one of the most common carnivores, although this classification is a bit of a misnomer as they readily consume anything edible. Well known for their ability to pry open garbage cans, raccoons will also feast on fruits or frogs, carrion or crayfish. Raccoons seem particularly attracted to water, which they play in and use to constantly clean their paws. Observing these curious and clever mammals is difficult to do by day, for they are primarily nocturnal.

**Striped skunk** (*Mephitis mephitis*). These primarily nocturnal scavengers wander Cold Canyon in search of berries, insects, eggs, and dead animals. Their characteristic black-

Virginia opossum

Striped skunk

Raccoon

Coyote

Bobcat

Mountain lion

and-white coloration warns most potential predators of their nauseating spray. As a result, skunks seem to bumble along their way without much concern for predators.

**Coyote** (*Canis latrans*). The coyote is a resident of Cold Canyon, but individuals are so secretive that few people get even a glimpse of them. These mammals, about the size of a collie, remain hidden by day but roam the homestead trail by night, sometimes barking and yipping in a cacophonous chorus. They leave behind a few obvious signs of their presence: their footprints (which are usually impossible to differentiate from the prints of other dogs) and their scats. The remains in their scats attest to their varied diet of berries, seeds, and small mammals.

**Bobcat** (*Lynx rufus*). These cats are a rare sight in Cold Canyon, but can be readily identified by their spotted coat, pointy ears, and stubby tail. Bobcats are true carnivores, chasing down mice, rabbits, birds, and snakes for their meals. Like many mammals, bobcats are primarily nocturnal, but they are also active in the mornings and late afternoons.

**Mountain lion** (*Puma concolor*). Perhaps the most famous of California's mammals, the mountain lion is also one of the rarest in the canyon. These large cats are active primarily at night, on the hunt for deer, rabbits, and other mammals, but on occasion they have been seen in the early morning and even the afternoon. By day, mountain lions generally retreat to the more remote parts of Wild Horse and Cold canyons. They are sometimes confused with bobcats, but their large thick tail and uniformly tawny fur easily distinguish them from Cold Canyon's other wild cat. The sight of a mountain lion is lucky indeed, but if you do see one, act threatening, make a lot of noise, and stand your ground to prevent the lion from treating you as potential prey.

**California ground squirrel** (*Spermophilus beecheyi*). Douglas California ground squirrels, the subspecies found in the reserve, are mottled squirrels with a dark blotch on their back and are one of the few mammals that are active by day. Although they occasionally climb into trees to find food or to sun themselves, they are most comfortable down near the ground, especially near their burrows. The high-pitched alarm whistle of a California ground squirrel is often mistaken for a bird call, especially because the squirrel is rarely in sight after it whistles. Ground squirrels have good reason to stay alert, for they are hunted by many other vertebrates in Cold Canyon, including birds of prey, foxes, and rattlesnakes. Of particular interest to some researchers at the University of California, Davis, are their interactions with rattlesnakes. Adult ground squirrels are immune to the venom of rattlesnakes, but their young are not. In the presence of a rattlesnake, adult ground squirrels are particularly aggressive and often attempt to drive the snake away from their vulnerable young.

**Deer mouse** (*Peromyscus maniculatus*). These are the most common mammals on the reserve, although their strictly nocturnal habits generally eliminate the possibility of seeing them. Deer mice have light brown fur on their backs, white fur on their bellies and legs, long tails, and large eyes and ears. By night they forage for seeds and berries, and will occasionally munch on insects. Deer mice are carriers of the Hanta virus, which has caused human deaths in the West. The possibility of an outbreak in Cold Canyon is remote, for

California vole

Dusky-footed woodrat

Deer mouse

Brush rabbit

California ground squirrel

Mule deer

even if there were diseased mice in Cold Canyon, most outbreaks of the disease have involved prolonged human contact with dried mouse feces.

**Dusky-footed woodrat** (*Neotoma fuscipes*). Dusky-footed woodrats are the architects of the mammalian world. They live in a complex structure built of sticks and other debris that can be up to 8 feet (2.4 m) in height and contain numerous tunnels and rooms. It is these houses that form the nucleus for the social life of woodrats. Older, dominant individuals usually maintain control over several houses. Except in the case of a mother woodrat with her young, individuals live alone in these houses. They shift from house to house, and occasionally move into new territory, taking over a group of houses from an individual that has died or moved on. The presence of surplus houses in an area is critical for the establishment of subadult woodrats dispersing from their mother's home.

Woodrats are generally nocturnal, usually requiring some disturbance to or near their house to cause them to emerge from the house during the daylight. Skilled climbers, woodrats are often found moving through the branches of trees once night approaches, when they forage for acorns, berries, seeds, and other vegetable matter. Woodrats may actually aid in the uphill dispersal of seeds as large as buckeyes: they carry these and other seeds to their middens, where they occasionally germinate.

**California vole** (*Microtus californicus*). This small, dark-gray mouse is most active by night, but may be heard at other times of the day as it travels along the runways it cuts through the grass. Voles can be distinguished from other mice by their short tails and hairy ears. They can be found wherever grass and other ground cover grows more than several inches high. Voles eat grass, as well as seeds from many plants.

**Brush rabbit** (*Sylvilagus bachmani*). These small brown rabbits with white tails are sometimes flushed from thick grass and shrubs. It is likely they have a permanent shelter site nearby, such as a burrow or a particularly tangled thicket, because they do not venture far from safety. Although they are active in the day, they are rarely conspicuous because the brush rabbit is prey to a great many animals.

**Mule deer** (*Odocoileus hemionus*). Deer are plentiful in the canyon, as attested by their often-sighted footprints and scats, but their fear of human hunters makes them a relatively uncommon sight. Deer often travel in small groups, and communicate with each other by stamping their feet. The best place to see deer is along the creek, near the less accessible pools. Look for the trails that they use to travel from the hill slopes to the pools.

# Hiking in the Reserve

Stebbins Cold Canyon Reserve is open for hiking year round, every day, from sunrise to sunset. Stebbins Cold Canyon Reserve is a 45-minute drive west of Davis and a two-hour drive northeast of San Francisco. From Davis, take Covell Blvd./E6 to Winters. From San Francisco, take Rte. 80 north to Rte. 505 north, and take the Winters exit off of 505. In Winters, take Hwy 128 towards Lake Berryessa. A silver fire gate will be obvious on the left, 0.2 miles past the point where Hwy 128 crosses Putah Creek. This is the main entrance to the reserve. A large parking area is located on the right side of the road just before you reach the small parking area at the gate.

## Reserve Rules

To keep the reserve as beautiful and as wild as you find it, please follow some simple rules when hiking in Cold Canyon.

1) Please sign in on the register located at the reserve entrance (about 200 m up the trail from the road). Knowing how many people use the reserve, and for what purpose, is crucial for proper reserve management.

2) Because Cold Canyon is a reserve, plants, animals, and other organisms on the reserve should not be collected except by explicit permission of natural reserve management staff. Contact reserve management for permission to collect organisms for scientific research.

3) Likewise, please do not disturb research sites or management areas. These areas will be indicated by colored tape and flags or by signs. Please obey any signs restricting access to trails that are hazardous or whose use is causing hillside erosion.

4) Minimize signs of your presence. Please pack out anything you bring in, especially trash. If you see trash left by other people, please pack that out as well.

5) Bikes, dogs, fires, and guns are not allowed on the reserve.

## The Trail Sytem in Cold Canyon

There are three main hiking trails in Cold Canyon, as shown on the trail map. **The Homestead Trail** starts at the main gate by the road and runs up the canyon to the old Vlahos homestead and Cold Storage Area. This one mile one-way trail is easy to moderate. **The Blue Ridge Loop Trail** begins 100 yards west of the main gate, on the west side of the creek. This trail takes you up to the western peaks overlooking Cold Canyon and Lake Berryessa. Views are spectacular, although the trail is steep. It proceeds along the ridge and loops back to the old stone Homestead foundation. The full loop is a challenging 5 miles. (The easier route starts at the Homestead and proceeds clockwise around the loop and descends on the Blue Ridge Loop.) **The Pleasants Ridge Trail** begins at the left fork just inside the main gate. This very steep and unmaintained trail runs up to the top of the eastern ridge. Use caution when climbing this one mile trail.

## Potential Hazards of Cold Canyon

Cold Canyon is a beautiful area, but it has its hazards. The trail has become quite rough and steep in places as a result of landslides, and creek crossings contain slippery rocks. Watch your step! In the summer, and even in the spring, Cold Canyon can get quite warm, so the best time to hike is during the cooler morning. If you plan to hike later in the day, bring a hat and water to prevent heat exhaustion during hiking.

Three species of vertebrates living on the reserve are potentially dangerous. Mountain lions, black bears, and rattlesnakes are present, although your probability of encountering these species is remote. With the possible exception of mountain lions, most animals will be more scared of you, and will not be a threat as long as you respect the animals' space. Black bears will most likely flee as soon as you encounter them. Rattlesnakes will usually warn you with the buzz of their rattle. Because mountain lions can be more dangerous than bears and snakes, always travel with a partner, and keep children in sight. In the very rare case that you encounter a moutain lion, act threatening: stand your ground, make noise, stand tall, and if necessary, throw sticks or rocks.

Other less hazardous dangers may be hidden in the trailside foliage. Poison oak (*Toxicodendron diversilobum*) is thick along the trail and elsewhere, so avoid brushing up against either its foliage or bare branches (p. 26). In addition, deer ticks are present. Deer ticks are potential carriers of Lyme disease, and the best way to avoid the disease is to avoid the ticks. Wear long pants to protect yourself from both poison oak and deer ticks. Pre-exposure cremes are available to reduce problems with poison oak, and insect repellent sprayed on your shoes, socks, and pants will further reduce your likelihood of picking up ticks.

**Leaflets three, let them be! Beware of poison oak.**

## A Natural History Hike for All Seasons

This tour describes sights along Cold Canyon's homestead trail. Use the accompanying map (inset of back cover) to locate these sights as you walk along the trail.

**1. Outlet of Cold Creek** (unmarked). Start at the parking area 0.2 miles past the bridge over Putah Creek on the right-hand side. In this area you can view the lowermost reaches of Cold Creek, which flows out of Stebbins Cold Canyon Reserve. You can leave your car in either this parking lot or the next pull-off on the left side of the road, at the reserve gate. In any case, take a moment to look at the mouth of Cold Creek.

The channel in front of you is an example of channelization that was done to enable water to move quickly through the area. This project was finished in the summer of 1995 in response to the problems created by the immense debris flows that came out the canyon in the previous winter, one of the wettest on record. You will see evidence of these flows when you hike up the canyon. In response to the rains, huge landslides occurred in the canyon and the material was flushed into Putah Creek, much of it winding up in Solano Reservoir. Landslides are natural events in this steep, geologically active landscape. Indeed, the soil on which regional agriculture depends arguably originated from such events accumulated over thousands of years (old maps mark the Davis region as "swamps, impassible in winter"). Unfortunately, humans can no longer allow the sediment to spread across the valley floor. Instead, it fills in reservoirs and otherwise becomes a problem for us.

Just upstream of the mouth of Cold Creek is a major spawning area for wild trout in Putah Creek. If you return to this spot in December and January, you may be able to see the shadowy forms of large trout spawning in the riffles. Their presence is marked by the distribution of fly fishers! Major efforts by anglers to improve the spawning habitat were largely negated by the floods of 1995, so the future of trout spawning in this region is not known.

**2. Roadcut** (unmarked). As you walk up the road to the trail, observe the exposure of bedrock on the slope to your left. This exposure consists of many layers of sandstone and is part of the Sites Formation, the bedrock forming the entire eastern slope of Cold Canyon. See if you can find the fault running through this exposure by following the layers across the exposure. The fault will appear as a discontinuity in the layering. For example, thin layers will suddenly change to thick layers. The fault marks a relatively small discontinuity (the bedrock on one side of the fault was moved upward only a few feet), but it extends up the entire slope. The existence of this fault hints at the possibility that the bedrock underlying Cold Canyon is still being moved about by geological forces, although at a very slow rate.

**Strata of the Sites Formation revealed in the road cut near the entrance to Cold Canyon.**

**3. Main gate**. This gate once blocked public entrance to a fire road which went a considerable distance up the canyon. The homestead trail is all that remains of the fire road. Just inside the gate, the trail forks. The left fork is the Pleasants Ridge trail, which winds up the hill through savanna dominated by blue oaks (*Quercus douglasii*), to the top of the eastern ridge. This trail is steep and not as well maintained as the homestead trail, so use caution if you choose to hike on this trail. The right fork marks the start of the homestead trail. Follow this trail toward the creek.

**4. Poison oak**. If you are not familiar with the way poison oak looks, use these plants as a standard. In addition to poison oak, look for the large elliptical leaves of mule-ears (*Wyethia helenioides*) growing to the left of the marker. Mule-ears sprout large sunflower-like blossoms in mid- to late spring. Later in the year, the plant may die back, but the skeletons of the dead leaves may persist through the winter.

**5. Western redbud** (*Cercis occidentalis*) starts the season with a gorgeous display of magenta blossoms, followed by its heart-shaped leaves. It often sheds these in the heat of the summer, so that by fall, its brown bean-shaped fruits may be all that remain on the branches.

**6. Landslide**. At the first bend are the remains of a major landslide caused by the heavy rains of the winter of 1995. The debris flowed out of the dry canyon on your left (as you face upstream). The debris flow eliminated much of the riparian vegetation in the creek below and nearly took out the highway as well. Remember that this was a natural event! As you examine the inundation on your left, look for pitcher sage (*Lepechinia calycina*), a shrub with large pink-lavender tubular blossoms that appear in April.

**7. Fork to Blue Ridge trail**. Just after you cross the slide, you will see a large rock on your left that is covered with small circular patches of greenish and rust-colored lichens. Lichens (a fungi-algae symbiotic organism) gradually dissolve rocks, creating soil, and are part of the whole process of constant change symbolized by the landslides. The trail to the right marks the start of the Blue Ridge trail. After the creek crossing, the trail winds uphill to the rocky ridge top.

**8. Interior live oak**. From the lichen rock, proceed along the trail across a boulder-strewn area, more remains of the 1995 landslide. In the midst of this area is a large interior live oak (*Quercus wislizeni*) that managed to survive the slide.

**9. Coyote brush**. Coyote brush (*Baccharis pilularis*) can be found in many spots along the trail. This species is notable because it is dioecious: each individual plant is either male or female, just like in humans. In contrast, most other plants are monoecious, and therefore produce both seeds and pollen on the same plant - usually within a single flower.

**10. Sign-in area**. This is the entrance to the UC reserve. **Please sign in** - an accurate count of users is vital to the reserve. The university has made a deliberate policy to keep this reserve (unlike many of the others in the UC Natural Reserve System) open to the public. To ensure that it remains open, please follow all reserve rules and encourage others to do the same.

**Kiosk, rules, and sign-in area.**

Just past the sign-in box, a toyon (*Heteromeles arbutifolia*) grows on the right side of the trail. If you visit in the fall or winter, its bright red fruit will be clearly visible.

**11. Slide area**. About 330 feet (100 m) up the trail is a flat open space that resulted from another slide, in this case a chunk of the hillside on the left. A side path on the right leads to a grassy area by Cold Creek, where Henderson's shooting star (*Dodecatheon hendersonii*) can be found growing in the early spring. Next to the trail at this point, there is a skunk-brush (*Rhus trilobata*). It is a close relative of poison oak and can easily be confuesed with it. It does not cause skin irritation.

**12. Manzanita**. The shrubs with pastel-green leaves and reddish, twisted branches are parry manzanita (*Arctostaphylos manzanita*). This plant is a dominant species in chaparral, along with chamise (*Adenostoma fasciculatum*) and toyon. In the fall, look up the left side of the trail for a beautiful display of this plant's bright red berries. Another common chaparral plant, mountain mahogany (*Cercocarpus betuloides*), grows on the left side of the trail, directly opposite the location marker.

**13. Yerba santa**. In the summer, yerba santa (*Eriodictyon californicum*) is almost more formidable looking than poison oak; its long dark oily leaves make it appear particularly noxious. While this oil may in fact ward off potential herbivores, it appears to have medicinal properties for humans. In the winter and spring, the leaves of this plant become less dark in color and much less oily.

**14. Bay trees**. Large bay trees (*Umbellularia californica*) stand above and to your left. These have relatively thin, straight limbs and light gray bark. Their leaves are narrow and smooth-edged, and give off a pleasant, pungent smell when rubbed.

**15. Rocky peak on right**. As the trail starts to climb, marked by wooden erosion control bars, note the rocky peak across to the right. Usually, you can see raptors soaring around it. Mostly they are Turkey Vultures (*Cathartes aura*), but Red-tailed Hawks (*Buteo jamaicensis*) are also common. Rarer are birds such as Golden Eagles (*Aquila chrysaetos*) or Peregrine Falcons (*Falco peregrinus*). If you want an ambitious hike, you can follow the Blue Ridge trail, which starts at the highway on the other side of Cold Creek and climbs to the top of the ridge. Views are spectacular – much of Lake Berryessa can be seen, and on clear days, Sacramento and the Sierra Nevada as well. The soaring birds will be below you.

In the moist areas along the left side of this section of trail, milk maids (*Cardamine californica*), delicate white four-petaled flowers, begin blooming starting in February. Look also for ferns and bushes of the bush monkeyflower (*Mimulus aurantiacus*). The bright orange-yellow tubular flowers that bloom in the spring make this one of the showiest chaparral shrubs.

**Bush monkeyflower.**

**16. Old landslide with buckeyes**. About 50 feet (15 m) ahead, when the trail starts to climb abruptly with some "stairs," you will know you are crossing a landslide that occurred in 1982. Actually, it was more of a land slump because a large chunk of the hillside slid downwards with much of the vegetation still on top. Much of the slope did lose its plant cover, but it is now difficult to tell this portion of the slump from the surrounding hillside.

In very early spring, shooting stars (*Dodecatheon hendersonii*) blooms along the bank here. Numerous California buckeyes (*Aesculus calfornica*) grow along this section of the trail. They can be identified by their smooth, pale gray bark. In spring, their large compound leaves make this area very lush. In the late spring, their showy flowers bloom – look for large candelabra of small white or pink flowers. Both flowers and leaves wither over the summer, revealing the developing fruits. These are the "buckeyes," huge light brown, shiny seeds, that remain covered by a wrinkled pear-shaped husk until late fall. When the fall rains begin, the husks open and drop the huge seeds onto the ground. Just south of the sign, a large California pipevine (*Aristolochia californica*) is climbing up a buckeye. Look for flowers in mid-February.

**17. Intermittent drainage**. You will soon come to a drainage that in the spring delivers water from the slopes on the left. Just before this drainage is an open area that will allow you to see the steep slope on the opposite side of the creek. A flow from a small landslide scars this slope. You can follow the scar up to the landslide scarp, the ragged wall of soil exposed by the earth that tumbled down the slope. This is the remains of a relatively small landslide. Thousands of years ago, a much larger landslide occurred, and the remains lie right above this small landslide. Look up to the ragged peak topping Blue Ridge. The sheer cliff forming this peak is the remains of the much older landslide scarp. The earth that was moved in this landslide now forms the entire hill slope below the cliff, extending both to the left and right of the cliff in a relatively flat line along the hill.

**Geological features on the hillslope: a land slide is at left and the spring is at lower right.**

Several other interesting geological features occur very close to the small landslide scar. Slightly below and to the right of the landslide scarp is a mineral spring. From the trail, you may be able to spot the spring on the slope, located above some tall reeds. Here, a small whitish column has formed from the precipitation of minerals dissolved in the spring water. The presence of the spring suggests that a large fault may run through Cold Canyon (p. 15).

Just downstream from the landslide scar, at the creek's edge, is a remarkable outcrop of bedrock. The bedrock is composed of many layers of sandstone, but these layers of rock have somehow been turned completely upside down from the position in which they originally lay millions of years ago. Read about how these rock layers could have been overturned in the geology section (p. 16).

As you cross the drainage, look for several plants of the bush lupine (*Lupinus albifrons*), a shrub with silvery-gray leaves and spikes of blue flowers that bloom mid-spring.

**18. Wild grapes**. Soon you will step into chaparral habitat, although the proximity of the creek provides moisture for other plants to grow. Among them is the California wild grape (*Vitis californica*), a vine that can be seen covering much of the vegetation on the right side of the trail. These plants produce their fruits in late summer. They are edible, but extremely seedy and sour compared to cultivated varieties of grapes. Nevertheless, these fruits provide a very important food source for birds and other animals.

In the fall and winter, this area is often used as a shelter spot for thousands of ladybugs (*Hippodamia convergens*). These friends to gardeners have a red covering with black spots, and they congregate in dense clusters throughout the winter. Why they form such huge clusters is not known.

Look past the wild grapes to the creek, where you can see young cottonwoods (*Populus fremontii*) and willows (*Salix* spp.) that rooted after the huge storm-generated rock slides and floods of 1995.

**19. Spice bush**. Several spice bush (*Calycanthus occidentalis*) grow here next to the location marker. True to their name, spice bush leaves are quite piquant smelling, and remind one of California bay, a close relative. In the spring, these shrubs produce large deep red flowers with many petals. By summer these petals have fallen off, leaving a bare oval seed capsule

Another showy plant – California fuchsia (*Epilobium canum*) – occurs about 10 feet down the side trail on the left. Its long red tubular flowers first appear in August. However, this plant may continue to bloom through November, making it one of the latest-blooming plants in Cold Canyon. Hummingbirds eager for nourishment in the fall flock to these plants to sip the flowers' nectar.

**20. Chaparral plants.** This section of the trail is bordered by extensive chaparral. These plants are extremely well adapted to fire - new branches and foliage sprout from the old rootstocks. Chief among these plants is chamise, the wiry bush with tiny leaves that dominates this area. A chamise plant occurs right above the location marker. The area in which you are standing burned in 1988 and resprouted. It is very difficult to see now that it burned at all.

**21. Side trail**. A scrub oak (*Quercus berberidifolia*) grows at a fork in the path. The left fork is the main trail, and a side trail on the right will take you toward the creek. Several interesting features can be found at the end of this short trail. In the spring, the flowing creek forms a pool here. This is a great place to look for aquatic insects, foothill yellow-legged frogs (*Rana boylei*) , and other animals. On the opposite side of the pool, observe the rock outcrops. Here, two types of rocks can be seen: on your left is the dark, soft, and thinly layered shale of the Yolo Formation. On the right is the tan-colored and more thickly layered sandstone of the Venado Formation.

A third interesting feature is located in a large boulder at the end of the trail. Look for a rounded hole in the rock. This is a mortar made by Native Americans several hundred to several thousand years ago. They used this mortar to grind acorns and other seeds into flour, which formed the staple of their diet.

**22. Cold Creek**. Return to the main trail. The trail next crosses the bed of Cold Creek. In most years, this stream flows for about seven months (November-May). When it starts flowing, it is quickly colonized by aquatic insects such as water striders (*Gerris remigis*) and whirligig beetles (*Gyrinus plicifer*), as well as stringy algae. Winter rains bring out California newts (*Taricha torosa*) to breed in the stream. In the summer, pools of water are left behind when the stream stops flowing and are important sources of water for California Quail (*Callipepla californica*), mule deer (*Odocoileus hemionus*), mountain lions (*Puma concolor*), and other wildlife.

**Creek crossing looking downstream, with mule-fat bushes on both sides.**

The creek crossing is also a good place to look at plants. The plants growing in and along the streamcourse are mule-fat (*Baccharis salicifolia*), which looks like willow, but is a completely unrelated species. In June through August, look for the tall stalks and blooms of the California sunflower (*Helianthus californicus*) along the edge the creek. On the steep left side of the creek, blue larkspur (*Delphinium hesperium*) makes a showy display in the spring. Just after the creek crossing, the vertical slope on the right yields a colorful display of red larkspur (*Delphinium nudicaule*) from March to early April.

If you are here in the spring, water is usually still flowing. Carefully cross the creek using the rocks that stick out of the water (the rocks can be slippery). In the summer and fall, the creek is usually dry, and crossing will not be difficult. If water is present, examine the pools and note what life is present.

On the left slope at the creek crossing is another bedrock outcrop, in this case a section of sandstone and mudstone from the Sites Formation forming the eastern slope.

**23. Gray or Foothill pine tree**. As you hike up the trail from the creek crossing, about 50 meters up the trail on your left is a tall gray or foothill pine (*Pinus sabiniana*), Cold Canyon's only species of conifer. These trees are often home to a colonies of small (0.25 in., 3-5 mm long) grey-black ants (*Liometopum luctuosum*). These ants make tunnels in dead portions of the tree, which they defend aggressively from ants and other animals. This species is a generalist scavenger and forms long foraging lines, maintained and reinforced by a chemical trail pheromone. This species is native to California, very unlike the related Argentine ant (*Linepithema humile*), a notorious introduced pest that causes much trouble for both native ants and human homeowners in the valley.

**Grey or Foothill Pine.**

**24. Bedrock underlying Cold Creek**. As the trail continues upstream, look for a side trail on the left leading down to the creek. At this point in the canyon, the shale bedrock underlying the canyon makes its appearance. In the spring, the gray layers of the Yolo Formation form the bottom of the pools in this section of the creek. In the summer and fall, these pools have dried up, allowing you to inspect the bedrock. Shale is easily eroded by the action of water, so outcrops of the Yolo Formation are difficult to find in Cold Canyon.

As you continue up the trail from here, look for large tufts of purple needlegrass (*Nassella pulchra*) and a wiry shrub called deerweed (*Lotus scoparius*) on the slope. Both are well-adapted to the hot dry summer climate of Cold Canyon. Purple needlegrass has roots extending deep into the ground to access water, while deerweed has small, sclerophyllous leaves. These plants can endure many summers of nasty heat, in contrast to fairy lanterns (*Calochortus amabilis*), small but beautiful yellow flowers that emerge in the spring, quickly flower and produce seeds, and then die. You may see some of these shy plants on this same slope if you look here in April or May.

**25. Cottonwood tree**. Continue up the main trail. As you cross the small bridge on the trail, you will have a good view of a cottonwood tree growing right in the stream channel. Cottonwoods are important riparian trees, requiring water around the roots. This is a good spot to watch birds, such as Black Phoebes (*Sayornis nigricans*) and Bewick's

Wrens (*Thyromanes bewickii*). In addition, this spot provides a good view of creekside flats that attract hundreds of butterflies searching for sunny spots on cool spring mornings. In certain years, the butterflies may blanket some areas like a colorful living carpet.

**A small mushroom.**

**26. Scrub oak**. The trail now climbs through chaparral, and this spot marks the location of a large scrub oak growing up and over you. The shade of the chaparral keeps this area relatively cool and moist, enabling maidenhair ferns (*Adiantum jordanii*) and mushrooms to thrive even after the spring rains have ceased.

**27. Tunnel of brush**. The chaparral now has become a veritable tunnel of buckbrush (*Ceanothus cuneatus*). As you pass through, towhees, wrens and small mammals can be heard skittering about just out of sight on either side of you.

**28. Wild Horse Canyon**. When you emerge from the chaparral, you will have a good view of both Cold Canyon close by and Wild Horse Canyon to the left in the distance. This wild area is owned partly by the University and partly by the Bureau of Land Management, with a few private inholdings. Notice the pattern of the vegetation, with grasslands on some slopes, chaparral on others. Overlooking these two drainages on Pleasants Ridge are the rocky outcrops of the Sites Formation.

**Save the hillslope! Stay on the trail.**

**29. No entry fence**. This fence was constructed to keep people from climbing this tempting open hillside. Even a small trail in this fragile land could start a major landslide. Look up the arroyo along the hillside and you can see where part of the hill slid down in 1995, creating the debris flow at your feet.

**30. Homestead**. A short distance further, the trail crosses the Vlahos homestead. The open area around you was probably cleared by Vlahos and the soil compacted by livestock, mainly goats. One indication of former grazing are the dry prickly plants occur among the grasses. These plants are yellow star-thistle (*Centaurea solstitialis*), a native of Asia that thrives on poor, over-grazed soils. The grassland is currently being treated to control this invasive pest, and with some luck and much hard work, we may be able to restore the native grassland vegetation that used to grow here.

**31. Homestead foundations**. The stone wall on the left is about all that remains of the Vlahos' former farmhouse. Walk around the foundations and you can find the former site of a well, some rusted metal, and a few rotted wood boards, but little else remains to remind us that this site was once a bustling household.

Beyond the homestead, the trail continues into a shady glen. In the spring, look for the large white and purple blossoms of the iris (*Iris macrosiphon*) and the faded-red flowers of the Indian warrior (*Pedicularis densiflora*) on the right side of the trail. Cross the creek to the foundations of the old cold storage shed and the end of the Homestead Trail.

**32. Cold storage area**. This area is delightfully cool and moist all summer long. The trees providing the shade include California bay and spice bush, but several large bigleaf maples (*Acer macrophyllum*) provide most of the shade. In the fall, the leaves of the maples turn color and bathe the glen in a wonderful orange light. In the spring, follow the creek up from the cold storage foundation to a small waterfall. Here Cold Creek pours over some boulders into a small pool that often contains a variety of aquatic wildlife.

The cold storage glen is one of the few areas of the reserve that was not burned by the 1988 fire. Most of the reserve only had the litter and plant understory burned by the fire, although a few trees were lost as well. The reserve was actually the site of backfires carefully set by the California Department of Fire to prevent further spread of the fire (which burned thousands of acres before it reached Cold Canyon) and to reduce the effects of fire on the reserve.

**The cold storage in summer.**

**The waterfall above the cold storage.**

# Appendices

The following appendices provide additional scientific information not covered in the previous chapters. Appendix 1 is a topographic map of the Cold Canyon region that includes known and inferred geographic features such as geologic formations, strike and dip measurements at particular outcrops, and faults. This information was collected as part of an undergraduate thesis by Courtney Ovqist.

Appendices 2 through 7 list the known plants and animals of Cold Canyon. Several caveats apply to these lists. First, each list must be considered incomplete, and new taxa will undoubtedly be found in the future. Second, as knowledge of different taxa vary quite a bit, not all the lists below are species lists. Most notably, the list of insect species would be an entire book in itself if all species were included, but unfortunately few families have been systematically documented in Cold Canyon. Therefore, the list of insects is subdivided by order and by family. Specific species lists of a few insect families can be found on the reserve's companion web site (**http://nrs.ucdavis.edu/stebbins.html**). Third, lists vary in the hierarchical breakdown of taxa. Categorizing the living world using the kingdom, phylum, class, order, family, genus, and species hierarchy is handy for biologists, but different taxa show characteristic differences at specific levels of this hierarchy. Therefore, plants are categorized based primarily on family, while other lists break species down initially by order. Fourth, as over 600 species are listed in the following pages, this brief field guide cannot effectively serve as an identification manual. Several references in the bibliography serve as good indentification guides. Where possible, we include some information on the biology of these different plants and animals. Further information on the more common species can be found in the appropriate sections of the guide.

# Appendix 1: Geologic and Topographic Map of Cold Canyon

**Map Symbols**

Contact
*Dashed where approximate*

Fault
*Dashed where approximate*

45
Anticline

60
Strike and Dip of Bedding

60
Stike and Dip of Overturned Bedding

Spring

Scarp and Landslide

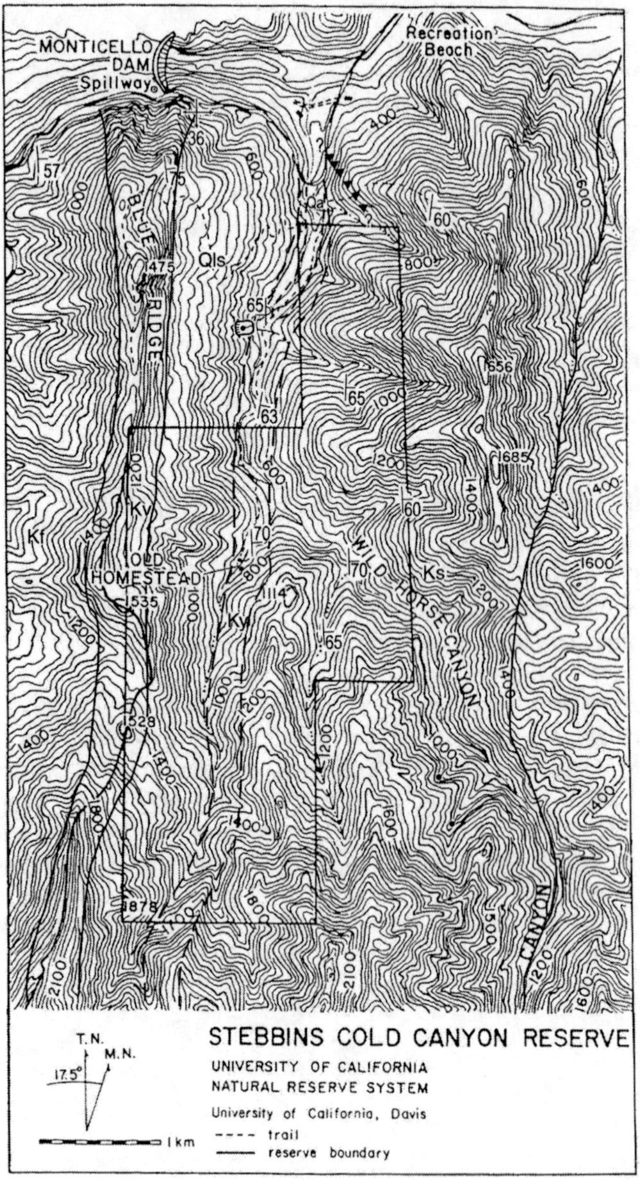


**Map Explanation**

| | | | |
|---|---|---|---|
| Quaternary | Qa | Aluvium | Stream gravel and boulders |
| | Qls | Landslide Deposits | Sandstone and shale colluvium |
| Cretaceous | Kfu | Funks Formation | Marine shale |
| | Ks | Sites Formation | Marine sandstone and shale |
| | Ky | Yolo Formation | Marine shale |
| | Kv | Venado Formation | Marine sandstone and shale |
| | Kfc | Fiske Creek Formation | Marine mudstone and sandstone |

# Appendix 2: Plant Species of Cold Canyon

| Scientific Name | Common Name | Habitat[13] | Blooms in |
|---|---|---|---|
| *FERNS AND THEIR ALLIES* | | | |
| **BLECHNACEAE** | **Chain Fern Family** | | |
| *Woodwardia fimbriata* | Giant chain fern | W L | |
| | | | |
| **DRYOPTERIDACEAE** | **Wood Fern Family** | | |
| *Dryopteris arguta* | Wood fern | S L | |
| | | | |
| **EQUISETACEAE** | **Horsetail Family** | | |
| *Equisetum arvense* | Common horsetail | M D | |
| *Equisetum hyemale* ssp. *affine* | Scouring rush | R | |
| | | | |
| **POLYPODIACEAE** | **Polypody Family** | | |
| *Polypodium californicum* | | L R | |
| | | | |
| **PTERIDACEAE** | **Maidenhair Fern Family** | | |
| *Adiantum jordanii* | Maidenhair fern | M L | |
| *Pellaea andromedifolia* | Coffee fern | X O | |
| *Pellaea mucronata* | Bird's-food fern | X O | |
| *Pentagramma triangularis* | Goldenback fern | O | |
| | | | |
| ***CONIFERS*** | | | |
| **PINACEAE** | **Pine Family** | | |
| *Pinus sabiniana* | Foothill or gray pine | S C L | Apr-May |
| | | | |
| ***FLOWERING PLANTS*** | | | |
| **DICOTS** | | | |
| **ACERACEAE** | **Maple Family** | | |
| *Acer macrophyllum* | Bigleaf maple | R | Apr-May |
| | | | |
| **ANACARDIACEAE** | **Sumac Family** | | |
| *Rhus trilobata* | Skunkbrush | X R L C | Mar-Apr |
| *Toxicodendron diversilobum* | Poison oak | Many | Apr-May |
| | | | |
| **APIACEAE** | **Celery Family** | | |
| *Angelica californica* | Angelica | X S G | June |
| *Anthriscus caucalis*[7] | Bur-chervil | D L | Apr-June |
| *Apiastrum angustifolium* | Wild celery | C | Mar-Apr |
| *Conium maculatum* | Poison hemlock | M D L | Apr-June |
| *Coriandrum sativum*[7a] | Coriander, cilantro | D | May-July |
| *Daucus pusillus* | Rattlesnake weed | X G | Apr-June |
| *Foeniculum vulgare*[7a] | Fennel | D | June-Oct |
| *Ligusticum apiifolium* | Celery-leaved lovage | G C L | June-July |

## APPENDIX 2: PLANTS

| | | | |
|---|---|---|---|
| *Lomatium californicum* | California rock parsnip | S C L | Apr-June |
| *Lomatium dasycarpum* | Lace parsnip | C S | Mar-June |
| *Lomatium macrocarpum* | Large lace parsnip | C S | Apr-June |
| *Lomatium utriculatum* | Bladder parsnip | G S | Feb-May |
| *Perideridia kelloggii* | Yampah | G | July-Aug |
| *Sanicula bipinnata* | Poison sanicle | G L | Apr-May |
| *Sanicula bipinnatifida* | Purple sanicle | G S | Mar-May |
| *Sanicula crassicaulis* | Gamble weed | G S | Mar-May |
| *Torilis arvensis*[7] | | D | May-June |
| *Torilis nodosa*[7] | Hedge parsley | D | Apr-June |
| *Yabea microcarpa* | | G C S | Mar-June |
| | | | |
| **ARISTOLOCHIACEAE** | **Pipevine Family** | | |
| *Aristolochia californica* | California pipevine | R C L | Jan-Apr |
| | | | |
| **ASCLEPIADACEAE** | **Milkweed Family** | | |
| *Asclepias cordifolia* | Purple milkweed | O C | May-July |
| *Asclepias fascicularis* | Narrow-leaved milkweed | X G | June-Sep |
| | | | |
| **ASTERACEAE** | **Sunflower Family** | | |
| *Achillea millefolium* | Yarrow | G | Mar-June |
| *Achyrachaena mollis* | Blowwives | G | Apr-May |
| *Agoseris grandiflora* | California dandelion | G C S | May-July |
| *Agoseris heterophylla* | Mountain dandelion | G | Mar-May |
| *Anthemis cotula* | Mayweed | D | May-July |
| *Artemisia douglasiana* | Mugwort | S L | June-Oct |
| *Aster radulinus* | Broad-leaved aster | X L | July-Oct |
| *Baccharis pilularis* | Coyote brush | R C | Aug-Dec |
| *Baccharis salicifolia* | Mule-fat | R | Mar-July |
| *Brickellia californica* | California brickellia | G | Aug-Oct |
| *Calycadenia ciliosa* | Rosin weed | X G | June-Sep |
| *Calycadenia pauciflora* | Small-flowered calycadenia | X G | July-Sep |
| *Carduus pycnocephalus*[7a] | Italian thistle | G | May-July |
| *Centaurea melitensis*[7a] | Tocalote | D G S | Apr-June |
| *Centaurea solstitialis*[7a] | Yellow star-thistle | D G S | May-Oct |
| *Chaenactis glabriuscula* var. *heterocarpha* | | G | Apr-June |
| *Chamomilla suaveolens*[12] | Pineapple weed | D G | Apr-Aug |
| *Conyza canadensis* | Horseweed | D | Sep-Oct |
| *Cotula australis*[11] | Brass buttons | D | Jan-May |
| *Ericameria arborescens* | Golden fleece | C S | Aug-Nov |
| *Erigeron foliosus* var. *hartwegii* | Narrow-leaved fleabane | G C | May-July |
| *Erigeron inornatus* | California rayless daisy | C O | Jun-Sep |
| *Eriophyllum lanatum* var. *achillaeoides* | Woolly sunflower | X O G R | May-July |

| | | | |
|---|---|---|---|
| *Euthamia occidentalis* | Western goldenrod | G | Sep-Nov |
| *Filago californica* | California filago | G | March-June |
| *Filago gallica*[7] | Narrow-leaved filago | X S | Apr-June |
| *Gnaphalium californicum* | Green cudweed | G | Jan-July |
| *Gnaphalium canescens* ssp. | Fragrance Everlasting | | July-Nov |
| *beneolens* | | D | |
| *Grindelia camporum* var. *cam-* | Great Valley grindelia | | Aug-Sep |
| *porum* | San Joaquin matchweed | O G | |
| *Gutierrezia californica* | Perennial Little Sunflower | G | Dec |
| *Helianthella californica* | California sunflower | R | Apr-June |
| *Helianthus californicus* | Field tarweed | G | June-Aug |
| *Hemizonia congesta* ssp. *luzu-* | | | Sep-Oct |
| *lifolia* | | D G | |
| *Hesperevax acaulis* | | G | |
| *Hesperevax sparsiflora* | White hawkweed | S X L | Mar-May |
| *Hieracium albiflorum* | Smooth cat's-ear | D | June-Aug |
| *Hypochaeris glabra*[7] | Prickly lettuce | O G | May-June |
| *Lactuca serriola*[7] | Common hareleaf | X | Sep-Oct |
| *Lagophylla ramosissima* | Common madia | G | May-Oct |
| *Madia elegans* ssp. *vernalis* | Threadstem madia | G S | Mar-June |
| *Madia exigua* | Slender tarweed | G S | May-July |
| *Madia gracilis* ssp. *gracilis* | Woolly malacothrix | D C | Apr-Aug |
| *Malacothrix floccifera* | Slender cottonweed | G | May-June |
| *Micropus californicus* | Slender woollyheads | X D G | Mar-June |
| *Psilocarphus tenellus* | California chicory | D C S | Apr-June |
| *Rafinesquia californica* | | G C | Apr-July |
| *Rigiopappus leptocladus* | Annual groundsel | D | Apr-June |
| *Senecio vulgaris*[7] | Milk thistle | D | Most months |
| *Silybum marianum*[7a] | Common soliva | D | May-July |
| *Soliva sessilis*[10] | Sow-thistle | D M | Mar-June |
| *Sonchus asper* ssp. *asper*[7] | Slender sowthistle | D | Most months |
| *Sonchus tenerrimus*[7a] | Twiggy wreath plant | X G | |
| *Stephanomeria virgata* | | G C S | July-Nov |
| *Uropappus lindleyi* | Foothill mule-ears | G C | Apr-June |
| *Wyethia helenioides* | | | Mar-Apr |
| | **Borage Family** | | |
| **BORAGINACEAE** | Fiddleneck | D G | |
| *Amsinckia menziesii* var. *inter-* | | | Mar-June |
| *media* | Flaccid cryptantha | X | |
| *Cryptantha flaccida* | Large-leaved hound's | C L | April |
| *Cynoglossum grande* | tongue | | Mar-June |
| | Little pectocarya | X G S | |
| *Pectocarya pusilla* | Valley popcorn flower | G S | April |
| *Plagiobothrys canescens* | Popcorn flower | G S | Mar-May |
| *Plagiobothrys nothofulvus* | Small popcorn flower | G L S | Mar-May |
| *Plagiobothrys tenellus* | | | Mar-May |

## APPENDIX 2: PLANTS

| **BRASSICACEAE** | **Mustard Family** | | |
|---|---|---|---|
| *Arabis modesta* | | X O | Mar-June |
| *Athysanus pusillus* | Sandweed | G O | Mar-June |
| *Barbarea orthoceras* | | R | May-Sep |
| *Barbarea vulgaris*[9] | Winter-cress | D | Apr-May |
| *Brassica juncea* | | D G | April |
| *Brassica nigra*[7] | Black mustard | D G | Apr-June |
| *Capsella bursa-pastoris*[7] | Shepherd's-purse | D | Year-round |
| *Cardamine californica* var. *integrifolia* | Milkmaids | L | Feb-Apr |
| *Cardamine oligosperma* | Annual bittercress | R | Mar-July |
| *Cardamine pennsylvanica* | Pennsylvania bittercress | R | May-June |
| *Cardaria draba*[7] | Hoary-cress | D | Mar-June |
| *Descurainia sophia*[9] | Tansy-mustard | D | |
| *Draba verna*[7] | Whitlow-grass | D | Feb-May |
| *Erysimum capitatum* | Wallflower | Many | Mar-July |
| *Guillenia lasiophylla* | California mustard | X D G | Mar-June |
| *Hirschfeldia incana*[9] | Short pod mustard | D R | May-Oct |
| *Lepidium strictum*[10] | Peppergrass | D | Mar-May |
| *Sinapis arvensis*[9] | Charlock | D | Mar-Oct |
| *Sisymbrium altissimum*[7] | Tumble-mustard | D G | May-July |
| *Sisymbrium officinale*[7] | Hedge-mustard | D | Apr-July |
| *Thysanocarpus curvipes* | Lace pod | R M S | Mar-May |
| *Tropidocarpum gracile* | Keel-pod | G | Mar-May |

| **CALYCANTHACEAE** | **Calycanthus Family** | | |
|---|---|---|---|
| *Calycanthus occidentalis* | Spice bush | R | Apr-Aug |

| **CAMPANULACEAE** | **Bellflower Family** | | |
|---|---|---|---|
| *Githopsis specularioides* | Common blue cup | C S | Apr-June |
| *Triodanis biflora* | Venus looking-glass | D | Apr-June |

| **CAPRIFOLIACEAE** | **Honeysuckle Family** | | |
|---|---|---|---|
| *Lonicera interrupta* | Chaparral honeysuckle | X G C | May-July |
| *Sambucus mexicana* | Blue elderberry | R | Mar-Sep |
| *Symphoricarpos mollis* | Trailing snowberry | G C O | Apr-June |

| **CARYOPHYLLACEAE** | **Pink Family** | | |
|---|---|---|---|
| *Cerastium glomeratum*[7] | Mouse-ear chickweed | X D G C | Feb-May |
| *Minuartia douglasii* | Douglas's Sandwort | C O | Apr-June |
| *Petrorhagia dubia*[7] | | D S | May |
| *Sagina apetala* | Dwarf pearlwort | D R | May |
| *Silene gallica*[7] | Common catchfly | D G | Feb-June |
| *Stellaria media*[7] | Common chickweed | D G S | Mar-June |
| *Stellaria nitens* | Shining chickweed | D R S | Mar-June |

| | | | |
|---|---|---|---|
| **CHENOPODIACEAE** | **Goosefoot Family** | | |
| *Chenopodium californicum* | Perennial goosefoot | G | Mar-June |
| **CISTACEAE** | **Rock-rose Family** | | |
| *Helianthemum scoparium* | Peak rush-rose | O | Mar-June |
| **CONVOLVULACEAE** | **Morning-glory Family** | | |
| *Calystegia occidentalis* | | X O C | May-June |
| *Calystegia purpurata* ssp. *purpurata* | Bindweed | C L | May-July |
| **CORNACEAE** | **Dogwood Family** | | |
| *Cornus glabrata* | Brown dogwood | L R | May-July |
| **CRASSULACEAE** | **Stonecrop Family** | | |
| *Crassula connata* | Sand pygmy | O | Feb-May |
| *Crassula tillaea*[1,7] | Mediterranean sand pygmy | O | Mar-April |
| *Dudleya cymosa* | Canyon live-forever | O | Apr-June |
| **CUCURBITACEAE** | **Gourd Family** | | |
| *Marah fabaceus* var. *agrestis* | Wild cucumber | R C | Feb-Apr |
| *Marah watsonii* | Watson's manroot | C | Mar-Apr |
| **CUSCUTACEAE** | **Dodder Family** | | |
| *Cuscuta californica*[2] | California dodder | C G | May-Aug |
| **DATISCACEAE** | **Datisca Family** | | |
| *Datisca glomerata* | Durango root | X R | May-July |
| **ERICACEAE** | **Heath Family** | | |
| *Arbutus menziesii* | Pacific madrone | L | Mar-May |
| *Arctostaphylos glandulosa* ssp. *glandulosa* | Eastwood manzanita | C | Jan-Apr |
| *Arctostaphylos manzanita* | Parry manzanita | C | Feb-Apr |
| **EUPHORBIACEAE** | **Spurge Family** | | |
| *Eremocarpus setigerus* | Turkey mullein | X G | May-Oct |
| *Euphorbia crenulata* | Chinese caps | X | Mar-Aug |
| *Euphorbia spathulata* | Spurge | D G | Mar-June |
| **FABACEAE** | **Pea Family** | | |
| *Astragalus gambelianus* | Gambel's dwarf locoweed | G C | Mar-June |
| *Cercis occidentalis* | Western redbud | C L R | Feb-Apr |
| *Hoita macrostachya* | Leather root | R | July-Aug |
| *Lathyrus vestitus* var. *vestitus* | Pacific peavine | C S | Apr-June |
| *Lotus crassifolius* | Buck lotus | D C L | May-Aug |

| | | | |
|---|---|---|---|
| *Lotus humistratus* | Hill lotus | G L | Mar-June |
| *Lotus micranthus* | Pink-flowered lotus | G | Mar-May |
| *Lotus purshianus* | Spanish lotus | D C | June |
| *Lotus scoparius* | Deerweed | D C | Mar-Aug |
| *Lotus wrangelianus* | Smooth hill lotus | D L | Mar-Sep |
| *Lupinus albifrons* | Bush lupine | D C | Mar-June |
| *Lupinus bicolor* | Miniature lupine | O C | Mar-June |
| *Lupinus microcarpus* var. *densiflorus* | White-whorled lupine | D G | Apr-June |
| *Lupinus succulentus* | Fleshy lupine | D G | Feb-May |
| *Medicago polymorpha*[7] | Bur clover | D | Mar-June |
| *Melilotus alba*[9] | White sweet clover | D G | May-Sep |
| *Melilotus indica*[9] | Yellow sweet clover | D G | Apr-Oct |
| *Pickeringia montana* | Chaparral pea | C S | Apr-May |
| *Rupertia physodes* | California tea | S | May-July |
| *Trifolium albopurpureum* | Rancheria clover | G C S | Mar-Apr |
| *Trifolium ciliolatum* | Tree clover | D G C | |
| *Trifolium gracilentum* | Slender clover | D G | April |
| *Trifolium hirtum*[9] | Rose clover | D | Mar-May |
| *Trifolium microcephalum* | Small-headed clover | M D R | March |
| *Trifolium oliganthum* | Few-flowered clover | L C | April |
| *Trifolium willdenovii* | Tomcat clover | M D G | Mar-May |
| *Vicia sativa*[7] | Common vetch | D | Apr-May |
| **FAGACEAE** | **Oak Family** | | |
| *Quercus berberidifolia* | Scrub oak | C S | April |
| *Quercus douglasi* | Blue oak | S | Mar-Apr |
| *Quercus lobata* | Valley oak | S | |
| *Quercus wislizeni* | Interior live oak | L | Mar-Apr |
| **GARRYACEAE** | **Silk Tassel Family** | | |
| *Garrya fremontii* | Silk tassel | C L | Feb-Mar |
| **GERANIACEAE** | **Geranium Family** | | |
| *Erodium botrys*[7] | Long-beaked filaree | X D | Mar-May |
| *Erodium brachycarpum*[7a] | | X D | Feb-May |
| *Erodium cicutarium*[7a] | Red-stemmed filaree | X D G C | Feb-May |
| *Erodium moschatum*[7a] | White-stemmed filaree | D | Feb-May |
| *Geranium dissectum* | Cut-leaved geranium | D | Mar-May |
| *Geranium molle*[7] | Annual cranesbill | D | May-July |
| **GROSSULARIACEAE** | **Gooseberry Family** | | |
| *Ribes malvaceum* | Chaparral currant | C L | Oct-Mar |
| *Ribes victoris* | Victor's gooseberry | C | Mar-Apr |
| **HIPPOCASTANACEAE** | **Buckeye Family** | | |
| *Aesculus californica* | California buckeye | R L | May-June |

| | | | |
|---|---|---|---|
| **HYDROPHYLLACEAE** | **Waterleaf Family** | | |
| *Emmenanthe penduliflora* | Whispering bells | X D | May-June |
| *Eriodictyon californicum* | Yerba santa | G S C | May-July |
| *Eucrypta chrysanthemifolia* | Common eucrypta | D C | Mar-June |
| *Nemophila heterophylla* | Vari-leaf nemophila | C S | Mar-July |
| *Nemophila pedunculata* | Meadow nemophila | G S R | Apr-Aug |
| *Phacelia cicutaria* | Caterpillar phacelia | G C L | Mar-May |
| *Phacelia imbricata* | Imbricate phacelia | G C S | Mar-June |
| *Phacelia tanacetifolia*[1] | Tansy phacelia | G | Mar-May |
| | | | |
| **HYPERICACEAE** | **St. John's Wort Family** | | |
| *Hypericum concinnum* | Goldwire | X | May-July |
| *Hypericum perforatum*[7] | Klamathweed | D G | June-July |
| | | | |
| **JUGLANDACEAE** | **Walnut Family** | | |
| *Juglans californica* var. *hindsii* | Black walnut | L | Apr-May |
| | | | |
| **LAMIACEAE** | **Mint Family** | | |
| *Lamium amplexicaule*[7] | Henbit | D G | Mar-Aug |
| *Lepechinia calycina* | Pitcher sage | G S | Apr-June |
| *Marrubium vulgare*[7] | Horehound | D | Mar-Aug |
| *Mentha pulegium* | Pennyroyal | M | June-July |
| *Mentha spicata*[7] | Spearmint | M | July-Oct |
| *Monardella villosa* | Coyote mint | O C S | June-Aug |
| *Salvia columbariae* | Chia | X D C | Mar-June |
| *Scutellaria tuberosa* | Tuberous skullcap | D C S | Mar-July |
| *Stachys ajugoides* var. *rigida* | Hedge nettle | M | July-Aug |
| | | | |
| **LAURACEAE** | **Laurel Family** | | |
| *Umbellularia californica* | California bay or laurel | M C L | Dec-May |
| | | | |
| **MALVACEAE** | **Mallow Family** | | |
| *Malva nicaeensis*[9] | Bull mallow | D | Mar-Sep |
| *Malva parviflora*[9] | Cheeseweed | D | Year-round |
| *Sidalcea diploscypha* | Fringed sidalcea | G S | Apr-June |
| | | | |
| **OLEACEAE** | **Olive Family** | | |
| *Fraxinus dipetala* | Flowering ash | S C L | Apr-May |
| | | | |
| **ONAGRACEAE** | **Evening Primrose Family** | | |
| *Clarkia affinis* | | C | Apr-May |
| *Clarkia concinna* | Red ribbons | S L | May-July |
| *Clarkia purpurea* ssp. *quadrivulnera* | Four-spot | C G S | Apr-July |
| *Clarkia unguiculata* | Elegant clarkia | X L C | May-June |
| *Epilobium brachycarpum* | Annual willow-herb | D G S | June-Sep |

| | | | |
|---|---|---|---|
| *Epilobium canum* ssp. *canum* | California fuchsia | O | Aug-Nov |
| *Epilobium minutum* | Chaparral willow-herb | D G S | Apr-Aug |
| | | | |
| **OROBANCHACEAE** | **Broomrape Family** | | |
| *Orobanche bulbosa*[3] | Chaparral broomrape | C | Apr-July |
| *Orobanche californica* | California broomrape | X O | May-June |
| *Orobanche fasciculata*[4] | Clustered broomrape | X O | Apr-July |
| *Orobanche uniflora*[5] | Naked broomrape | M | Mar-May |
| | | | |
| **PAPAVERACEAE** | **Poppy Family** | | |
| *Eschscholzia caespitosa* | Annual California poppy | C | Mar-June |
| *Eschscholzia californica* | California golden poppy | G S | Feb-Sep |
| | | | |
| **PHILADELPHACEAE** | **Mock Orange Family** | | |
| *Whipplea modesta* | Yerba de selva | L | Apr-June |
| | | | |
| **PLANTAGINACEAE** | **Plantain Family** | | |
| *Plantago erecta* | Annual plantain | G S | April |
| | | | |
| **POLEMONIACEAE** | **Phlox Family** | | |
| *Allophyllum divaricatum* | Pink false gilia | D C S | Apr-June |
| *Allophyllum gilioides* | Blue false gilia | W G | Apr-June |
| *Collomia heterophylla* | Vari-leaf collomia | G | Apr-June |
| *Gilia capitata* | Blue-headed gilia | X G | May-July |
| *Gilia clivorum* | Many-stemmed gilia | G | Mar-Apr |
| *Gilia tricolor* | Bird's eyes | G | Mar-Apr |
| *Linanthus androsaceus* | False baby stars | C S L | Apr-June |
| *Linanthus bicolor* | True baby stars | GS | Mar-June |
| *Linanthus ciliatus* | Whisker brush | X G S | Mar-July |
| *Navarretia mellita* | Honey-scented navarretia | G | July |
| *Navarretia pubescens* | Downy navarretia | G | May-July |
| *Navarretia viscidula* | Sticky navarretia | W G | May-July |
| *Phlox gracilis* | Slender phlox | X-M | Mar-Aug |
| | | | |
| **POLYGONACEAE** | **Buckwheat Family** | | |
| *Chorizanthe membranacea* | Pink spineflower | G S C | Apr-July |
| *Eriogonum dasyanthemum* | Many-flowered buckwheat | X O | June-Oct |
| *Eriogonum nudum* | Naked buckwheat | X O | June-Nov |
| *Pterostegia drymarioides* | | M L | Mar-July |
| *Rumex acetosella*[9] | Sheep sorrel | D M | Mar-Aug |
| *Rumex crispus*[9] | Curly dock | D | Year-round |
| | | | |
| **PORTULACACEAE** | **Purslane Family** | | |
| *Calandrinia breweri* | Brewer's calandrinia | D | Mar-June |

| | | | |
|---|---|---|---|
| *Calandrinia ciliata* | Red maids | G | Feb-May |
| *Claytonia parviflora* ssp. *parviflora* | Spring beauty | D M | March |
| *Claytonia perfoliata* | Miner's lettuce | D G S L | Feb-May |
| | | | |
| **PRIMULACEAE** | **Primrose Family** | | |
| *Anagallis arvensis*[7] | Scarlet pimpernel | D | Mar-July |
| *Dodecatheon hendersonii* | Henderson's shooting star | L | Feb-May |
| | | | |
| **RANUNCULACEAE** | **Buttercup Family** | | |
| *Clematis lasiantha* | Virgin's bower | C S | Mar-May |
| *Delphinium hesperium* ssp. *pallescens* | Blue larkspur | S L | Apr-June |
| *Delphinium nudicaule* | Red larkspur | L | Mar-June |
| *Delphinium patens* | Zigzag delphinium | G S | Mar-May |
| *Ranunculus hebecarpus* | Pubescent-fruited buttercup | L C | Mar-May |
| *Ranunculus occidentalis* | Western buttercup | G S | Mar-May |
| | | | |
| **RHAMNACEAE** | **Buckthorn Family** | | |
| *Ceanothus cuneatus* var. *cuneatus* | Buck brush | C | Mar-May |
| *Ceanothus oliganthus* var. *sorediatus* | Jim brush | C | Feb-May |
| *Rhamnus californica* | California coffeeberry | C S | May-July |
| *Rhamnus crocea* | Redberry | C S | Mar-Apr |
| | | | |
| **ROSACEAE** | **Rose Family** | | |
| *Adenostoma fasciculatum* | Chamise | C | May-June |
| *Aphanes occidentalis* | Western lady's mantle | M G C | Mar-June |
| *Cercocarpus betuloides* | Mountain mahogany | C L | Mar-May |
| *Heteromeles arbutifolia* | Toyon | S C L | Apr-July |
| *Holodiscus discolor* | Ocean spray | G S D | May-Aug |
| *Potentilla glandulosa* | Sticky cinquefoil | Many | May-July |
| *Rosa californica* | California wild rose | R L | May-Aug |
| *Rosa gymnocarpa* | Wood rose | L | Mar-July |
| *Rubus ursinus* | California blackberry | R | May |
| | | | |
| **RUBIACEAE** | **Madder Family** | | |
| *Galium aparine*[7] | Cleavers | G S | Mar-July |
| *Galium bolanderi* | Bolander's bedstraw | C | May-Aug |
| *Galium murale*[7] | Tiny bedstraw | M G S | Apr-May |
| *Galium parisiense*[7] | Wall bedstraw | X G O | Apr-Aug |
| *Galium porrigens* | Climbing bedstraw | C | June |
| | | | |
| **SALICACEAE** | **Willow Family** | | |
| *Populus fremontii* ssp. *fremontii* | Fremont cottonwood | R | Mar-Apr |

| | | | |
|---|---|---|---|
| *Salix exigua* | Sandbar willow | R | Mar-May |
| *Salix laevigata* | Red willow | R | Mar-May |
| *Salix lasiolepis* | Arroyo willow | R | Feb-Apr |
| *Salix melanopsis* | Dusky willow | R | Mar-May |
| **SAXIFRAGACEAE** | **Saxifrage Family** | | |
| *Lithophragma heterophyllum* | Hill star | L | Mar-June |
| *Saxifraga californica* | California saxifrage | L | Feb-June |
| **SCROPHULARIACEAE** | **Figwort Family** | | |
| *Antirrhinum vexillo-calyculatum* | Wiry snapdragon | D O | May-Nov |
| *Castilleja affinis* | Coast paintbrush | X C | Apr-May |
| *Castilleja applegatei* ssp. *martinii* | Martin's Indian paint-brush | X C | May-Sep |
| *Castilleja attenuata* | Valley tassels | G | Mar-May |
| *Castilleja exserta* | Purple owl's-clover | G | Mar-May |
| *Castilleja foliolosa* | Woolly paintbrush | X O C | Mar-June |
| *Collinsia heterophylla* | Chinese houses | L S | Mar-June |
| *Collinsia sparsiflora* | Spinster's blue-eyed Mary | D G C | Mar-May |
| *Keckiella lemmonii* | Lemmon's penstemon | C S L | June-Aug |
| *Mimulus auriantiacus* | Bush monkeyflower | D C S | Mar-Aug |
| *Mimulus cardinalis* | Scarlet monkeyflower | R | Apr-Oct |
| *Mimulus guttatus* | Seepspring monkey-flower | R | Mar-Aug |
| *Pedicularis densiflora* | Indian warrior | C L | Jan-June |
| *Penstemon heterophyllus* | Foothill penstemon | G C | Apr-July |
| *Scrophularia californica* | California figwort | M C | Feb-July |
| *Tonella tenella* | Small-flowered tonella | L | Mar-May |
| *Triphysaria pusilla* | Dwarf orthocarpus | G | Apr-May |
| *Veronica persica*[7] | Persian speedwell | D G | Feb-May |
| **SOLANACEAE** | **Nightshade Family** | | |
| *Solanum xanti* | Purple nightshade | C L | Feb-July |
| **VALERIANACEAE** | **Valerian Family** | | |
| *Plectritis ciliosa* | Long-spurred plectritis | S | Mar-May |
| *Plectritis congesta* | Short-spurred plectritis | S | Apr-June |
| *Plectritis macrocera* | White plectritis | S | Apr-May |
| **VITACEAE** | **Grape Family** | | |
| *Vitis californica* | California wild grape | R | May-June |
| **MONOCOTYLEDONS** | | | |
| **CYPERACEAE** | **Sedge Family** | | |
| *Carex nudata* | Torrent sedge | R | |
| *Cyperus eragrostis* | Umbrella sedge | R | May-Nov |

| | | | |
|---|---|---|---|
| **IRIDACEAE** | **Iris Family** | | |
| *Iris fernaldii*[6] | Fernald's iris | L | Apr-May |
| *Iris macrosiphon* | Long-tubed iris | L S | Apr-May |
| *Sisyrinchium bellum* | Blue-eyed grass | G S | Mar-May |
| | | | |
| **JUNCACEAE** | **Rush Family** | | |
| *Juncus balticus* | Baltic rush | L R | May-Aug |
| *Juncus bufonius* | Toad rush | D R | Apr-Sep |
| *Juncus mexicanus* | Mexican rush | R | May-Aug |
| *Juncus patens* | Spreading rush | M G R | June-July |
| *Juncus xiphioides* | Sword-leaved rush | M G R | May-Oct |
| | | | |
| **LILIACEAE** | **Lily Family** | | |
| *Allium serra* | Wild onion | G | Mar-May |
| *Brodiaea elegans* ssp. *elegans* | Elegant brodiaea | G S | May-June |
| *Calochortus amabilis* | Golden fairy lantern | S L C | Apr-June |
| *Calochortus luteus* | Gold nuggets | G S | May-June |
| *Calochortus splendens* | Splended mariposa lilly | C G | May-June |
| *Chlorogalum pomeridianum* | Soap plant | G S | May-Aug |
| *Dichelostemma capitatum* | Blue dicks | S C G | Mar-May |
| *Dichelostemma volubile* | Twining lily | S C | |
| *Fritillaria affinis* | Checker-lily | G C S | Feb-May |
| *Triteleia laxa* | Ithuriel's spear | G S C | Mar-June |
| *Triteleia lugens* | | G S C | Apr-May |
| *Zigadenus fremontii* | Death camas | G S | Mar-May |
| | | | |
| **ORCHIDACEAE** | **Orchid Family** | | |
| *Piperia elegans* | Rein-orchid | X G C | May-Sep |
| | | | |
| **POACEAE** | **Grass Family** | | |
| *Agrostis exarata* | Pacific bent grass | M D G | June-Aug |
| *Aira caryophyllea*[7] | Hairgrass | D G | Apr-June |
| *Avena barbata*[7] | Slender wild oat | D G | Mar-June |
| *Avena fatua*[7] | Broad-leaved wild oat | D G | Apr-June |
| *Brachypodium distachyon*[9] | Purple false brome | D G | Apr-May |
| *Briza minor*[7] | Little quaking grass | M S G | Apr-June |
| *Bromus carinatus* | California brome | S C L | Apr-Aug |
| *Bromus diandrus*[7] | Ripgut grass | D G | Apr-June |
| *Bromus hordeaceus*[7] | Soft chess | D G | Apr-July |
| *Bromus laevipes* | Smooth-stemmed brome | S L | May-Aug |
| *Bromus madritensis*[7] | Foxtail chess | D G S | Apr-June |
| *Bromus madritensis* ssp. *rubens*[7a] | Red brome | D G S | Mar-June |
| *Bromus sterilis*[7] | Poverty brome | D G | Apr-June |
| *Cynodon dactylon*[8] | Bermuda grass | D G | Mar-May |
| *Cynosurus echinatus*[7] | Hedgehog dogtail | D G | Apr-June |
| *Elymus glaucus* | Woodland wild rye | G C S | June-Aug |

| | | | |
|---|---|---|---|
| *Elymus trachycaulis* | Slender wheat grass | L M S | |
| *Gastridium ventricosum*[7] | Nit grass | D G | May-Sep |
| *Hordeum marinum* ssp. *gussoneanum*[7] | Mediterranian barley | D G S | Apr-June |
| *Hordeum marinum* ssp. *leporinum*[7] | Farmer's foxtail | M S G | Apr-June |
| *Lolium multiflorum*[7] | Italian ryegrass | D G | Mar-May |
| *Lolium temulentum*[7a] | Darnel | D G | Apr-June |
| *Melica californica* | California melic grass | G S | Apr-May |
| *Melica torreyana* | Small-flowered melic grass | C S | Mar-June |
| *Nassella lepida* | Foothill needlegrass | X G S L C | Mar-May |
| *Nassella pulchra* | Purple needlegrass | G C S | Mar-May |
| *Poa annua*[7] | Annual bluegrass | M D G | Jan-July |
| *Poa bulbosa*[7] | Bulbous bluegrass | D G | Apr-July |
| *Poa secunda* ssp. *secunda* | One-sided bluegrass | Many | Feb-June |
| *Polypogon maritimus*[7] | Mediterranean beard grass | M | June |
| *Polypogon monspeliensis*[7] | Beard grass | M | Apr-Aug |
| *Vulpia bromoides*[7] | Brome fescue | D G S | Apr-June |
| *Vulpia microstachys* var. *ciliatus* | Grey fescue | D | Mar-June |
| *Vulpia microstachys* var. *confusa* | Small fescue | X G | Mar-June |
| *Vulpia microstachys* var. *microstachys* | Stiff annual fescue | X G S | Apr-June |
| *Vulpia microstachys* var. *pauciflora* | Foxtail fescue | X G S<br>X G S | Apr-June |
| *Vulpia myuros* var. *hirsuta*[7] | Rattail fescue | X G | Mar-June |
| *Vulpia myuros* var. *myuros*[7] | | X G S | Mar-May |
| | | | |
| **TYPHACEAE** | **Cattail Family** | | |
| *Typha angustifolia* | Narrow-leaved cattail | R | June |

## Footnotes

(1) Rare

(2) Parasitic on many herbaceous and woody plants

(3) Parasite of Adenostoma fasciculatum

(4) Occasional parasite of Artemisia, Eriogonum, Eriodictyon, etc.

(5) Parasite of Saxifraga

(6) Apparent hybrid with Iris macrosiphon, may be only a local form of latter

(7) Introduced: native to Europe

(7a) Introduced: native to Southern Europe

(8) Introduced: native to Africa

(9) Introduced: native to Eurasia

(10) Introduced: native to South America

(11) Introduced: native to Australia

(12) Introduced: native to NW North America and NE Asia

(13) **Habitat Abbreviations**

- **C** - Chaparral
- **D** - Disturbed sites
- **G** - Grassland
- **L** - Live oak woodland
- **M** - Moist sites
- **O** - Rock outcrops
- **R** - Riparian areas
- **S** - Savanna
- **X** - Xeric (dry) sites

# Appendix 3: Spider Families of Cold Canyon

| Family | Common name | Habitat type | | |
|---|---|---|---|---|
| | | Chaparral | Woodland | Riparian |
| Nemesiidae | False tarantula | | X | |
| Agelenidae | Funnel-web spiders | X | X | |
| Araneidae | Orb-weaving spiders | | X | X |
| Amaurobiidae | Cribellate spiders | | X | |
| Clubionidae | | | | X |
| Corrinidae | | | X | |
| Dictynidae | Cribellate spiders | X | X | X |
| Gnaphosidae | Ground spiders | X | X | |
| Heteropodidae | Giant crab spiders | | X | X |
| Linyphiidae | Money spiders | X | X | X |
| Lycosidae | Wolf spiders | X | X | X |
| Miturgidae | | | X | X |
| Oecobiidae | | X | X | |
| Oxyopidae | | X | | |
| Philodromidae | | | X | X |
| Pholcidae | Social spiders | X | X | |
| Salticidae | Jumping spiders | X | X | X |
| Tengellidae | | X | X | |
| Tetragnathidae | Long-jawed orb weavers | | | X |
| Theridiidae | Comb-footed/widow spiders | | X | X |
| Thomisidae | Crab spiders | | X | X |

# Appendix 4: Insect Families of Cold Canyon[1]

| ORDER/Family | Common Name | Feeding Style[2] Larval | Feeding Style[2] Adult | Habitat[3] |
|---|---|---|---|---|
| *Incomplete Metamorphosis* | | | | |
| **THYSANURA** | **Silverfish** | | | |
| Lepismatidae | -- | S | S | decaying matter |
| **MICROCORYPHIA** | **Bristletails** | | | |
| Machilidae | -- | S | S | decaying matter, rocks |
| Meinertellidae | -- | S | S | decaying matter, rocks |
| **EPHEMEROPTERA** | **Mayflies** | | | |
| Baetidae | -- | H | H | *water*; midair |
| Leptophlebiidae | -- | | | *still water*; midair |
| Siphlonuridae | -- | P | P | *riffles*; midair |
| **ODONATA** | **Dragonflies/Damselflies** | | | |
| Aeschnidae | Darners | P | P | *submerged veg.*; midair |
| Calopterygidae | Broad-winged damselflies | P | P | *submerged veg.*; midair |
| Coenagrionidae | Narrow-winged damselflies | P | P | *submerged veg.*; midair |
| Libellulidae | Skimmers | P | P | *submerged veg.*; midair |
| **ORTHOPTERA** | **Grasshoppers/Crickets** | | | |
| Acrididae | Grasshoppers | H | H | vegetation |
| Gryllacrididae | Ground and camel crickets | H | H | vegetation, soil |
| Gryllidae | Crickets | H | H | soil, vegetation |
| Tetrigidae | Pygmy grasshoppers | | | vegetation |
| Tettigoniidae | Katydids | H | H | vegetation |
| **BLATTARIA** | **Cockroaches** | | | |
| Blattellidae | Wood cockroaches | O | O | decaying veg. |
| **MANTODEA** | **Mantises** | | | |
| Mantidae | -- | P | P | vegetation |
| **PHASMIDA** | **Walking sticks** | | | |
| Timemidae | Timema walking sticks | H | H | vegetation |
| **DERMAPTERA** | **Earwigs** | | | |
| Forficulidae | European earwigs | O | O | moist, dark places |
| **ISOPTERA** | **Termites** | | | |
| Hodotermitidae | Dampwood termites | W | W | pine trees, logs |
| Rhinotermitidae | -- | W | W | buried wood |

| | | | | |
|---|---|---|---|---|
| **EMBIOPTERA** | **Webspinners** | | | |
| Oligotomidae | -- | S | S | under rocks, leaf litter |
| | | | | |
| **PLECOPTERA** | **Stoneflies** | | | |
| Chloroperlidae | Green stoneflies | H | H | *water*; riparian veg. |
| Nemouridae | Spring stoneflies | P | H | *water*; riparian veg. |
| Perlodidae | Perlodid stoneflies | | | *moving water*; riparian veg. |
| | | | | |
| **PSOCOPTERA** | **Barklice** | | | |
| Psocidae | -- | O | O | vegetation, under rocks |
| | | | | |
| **HEMIPTERA** | **True bugs** | | | |
| Alydidae | Broad-headed bugs | H | H | vegetation |
| Anthocoridae | Minute pirate bugs | P | P | vegetation |
| Aradidae | Flat bugs | S | S | tree bark |
| Belostomatidae | Giant water bugs | P | P | open water, submerged veg. |
| Berytidae | Stilt bugs | H | H | vegetation |
| Coreidae | Squash bugs | H | H | vegetation |
| Corixidae | Water boatmen | O | O | water |
| Enicocephalidae | Gnat bugs | P | P | under rocks or debris |
| Gelastocoridae | Toad bugs | P | P | streamside |
| Gerridae | Water striders | P | P | still water |
| Largidae | Largid bugs | H | H | vegetation |
| Lygaeidae | Seed bugs | H | H | vegetation |
| Miridae | Leaf bugs, plant bugs | H | P | vegetation |
| Nabidae | Damsel bugs | P | P | vegetation |
| Notonectidae | Backswimmers | P | P | open water |
| Pentatomidae | Stink bugs | H | P | vegetation |
| Pyrrhocoridae | Red bugs, cotton stainers | H | H | vegetation |
| Reduviidae | Assassin bugs | P | P | flowers, vegetation |
| Rhopalidae | Scentless plant bugs | H | H | vegetation |
| Saldidae | Shore bugs | P | S | streamsides, vegetation |
| Scutelleridae | Shield bugs | H | H | vegetation |
| Tingidae | Lace bugs | H | H | vegetation, leaf undersides |
| | | | | |
| *Intermediate Metamorphosis* | | | | |
| **HOMOPTERA** | **Aphids and others** | | | |
| Achilidae | Achilid planthoppers | H | H | vegetation |
| Aleyrodidae | White flies | H | H | vegetation |
| Aphididae | Aphids | H | H | vegetation |
| Cercopidae | Spittle bugs, froghoppers | H | H | vegetation |
| Cicadellidae | Leafhoppers | H | H | *in vegetation*; on vegetation |
| Cicadidae | Cicadas | H | H | *roots*; vegetation |
| Delphacidae | Delphacid planthoppers | H | H | vegetation |
| Membracidae | Treehoppers | H | H | vegetation |
| Psyllidae | Psyllids, jumping plant lice | H | H | vegetation |

## APPENDIX 4: INSECTS

| | | | | |
|---|---|---|---|---|
| **THYSANOPTERA** | **Thrips** | | | |
| Aeolothripidae | Broad-winged thrips | P | P | vegetation, esp. flowers |
| Thripidae | Common thrips | H | H | vegetation |
| *Complete Metamorphosis* | | | | |
| **NEUROPTERA** | **Lacewings and others** | | | |
| Berothidae | Beaded lacewings | P | H P | midair |
| Chrysopidae | Green lacewings | P | P | vegetation |
| Corydalidae | Dobsonflies | P | P | *water*; riparian veg. |
| Hemerobiidae | Brown lacewings | P | P | vegetation, midair |
| Myrmeleontidae | Antlions | P | P | *sandy soil*; midair |
| Rhaphidiidae | Snakeflies | P | P | vegetation |
| Sialidae | Alderflies | P | P | *riffles*; riparian veg. |
| **MECOPTERA** | **Scorpionflies** | | | |
| Bittacidae | Hangingflies | P | P | vegetation |
| **COLEOPTERA** | **Beetles** | | | |
| Alleculidae | Comb-clawed beetles | O | H | *soil*; decaying matter |
| Anobiidae | Furniture beetles | W | H | wood, decaying matter |
| Anthicidae | Antlike flower beetles | O | H | *soil*; decaying matter |
| Bruchidae | Seed beetles | H | | *seeds*; vegetation |
| Buprestidae | Metallic wood-borers | W H | W N | *vegetation*; fallen wood |
| Byturidae | Fruitworm beetles | H | | *vegetation* |
| Cantharidae | Soldier beetles | P | H P | on vegetation |
| Carabidae | Predaceous ground beetles | P | P | *decaying matter*; rocks |
| Cerambycidae | Longhorn beetles | W | H | *wood*; vegetation |
| Chrysomelidae | Leaf beetles | H | H | leaves |
| Cicindelidae | Tiger beetles | P | P | *burrows*; on ground |
| Cleridae | Checkered beetles | P | H | *fallen wood*; vegetation |
| Coccinellidae | Ladybugs | P | P | vegetation |
| Cucujidae | Flat bark beetles | S | S | bark, decaying matter |
| Cupedidae | Reticulated beetles | | | under bark |
| Curculionidae | Weevils | H | H | vegetation |
| Dascillidae | Soft-bodied plant beetles | H | H | riparian vegetation |
| Dermestidae | Hide beetles | S | H | *carrion*; flowers |
| Dytiscidae | Predaceous water beetles | P | P | *water*; open water, air |
| Elateridae | Click beetles | O | H | *soil, decaying matter*; flowers, fallen wood |
| Erotylidae | Pleasing fungus beetles | S | S | under bark, fungi |
| Gyrinidae | Whirligig beetles | P | P | *water*; water surfaces |
| Heteroceridae | Mud-loving beetles | | | riparian sediments |
| Histeridae | Hister beetles | P | P | decaying matter |
| Hydrophilidae | Scavenger water beetles | P | S P | still water, decaying matter, soil |
| Lampyridae | Fireflies | P | | *ground*; vegetation |

| | | | | |
|---|---|---|---|---|
| Lathridiidae | Brown scavenger beetles | S | S | decaying matter. flowers |
| Leiodidae | Round fungus beetles | | | decaying matter |
| Meloidae | Blister beetles | P | H | *bee's nests, egg masses*; vegetation |
| Melyridae | Soft-winged flower beetles | P | P | flowers |
| Mordellidae | Tumbling flower beetles | H P | N | *vegetation*; flowers |
| Oedemeridae | False blister beetles | W | | *wood*; flowers |
| Pedilidae | Antlike flower beetles | O | H | *soil*; vegetation, decaying matter |
| Phalacridae | Shining flower beetles | S | | *fungi*; flowers |
| Rhizophagidae | Root-eating beetles | P | P | bark, ant galleries |
| Scarabaeidae | Scarabs | S H | S H | *burrows*; vegetation, dung, carrion |
| Scirtidae | Marsh beetles | | | *water*; vegetation, decaying matter |
| Scolytidae | Bark beetles | H | | bark, wood |
| Silphidae | Carrion beetles | S H O | S H O | carrion |
| Staphylinidae | Rove beetles | P Pa S | Pa S | decaying matter |
| Tenebrionidae | Darkling ground beetles | H | H | decaying matter |
| Trogositidae | Bark-gnawing beetles | P O | | decaying matter |
| | | | | |
| **TRICHOPTERA** | **Caddisflies** | | | |
| Brachycentridae | -- | H P | | *water* |
| Hydropsychidae | Web-spinning caddisflies | O | | *water*; riparian veg. |
| | | | | |
| **LEPIDOPTERA** | **Butterflies and moths** | | | |
| Adelidae | Fairy moths | H | N | *vegetation*; flowers |
| Arctiidae | Tiger moths | H | N | vegetation |
| Coleophoridae | Casebearers | H | | *vegetation* |
| Danaidae | Milkweed butterflies | H | N | *vegetation*; flowers |
| Gelechiidae | Gelechiid moths | H | | *vegetation* |
| Geometridae | Geometrid moths | H | N | *vegetation*; flowers |
| Hesperiidae | Skippers | H | N | *vegetation*; flowers |
| Incurvariidae | Incurvariid moths | H | N | *vegetation*; flowers |
| Lycaenidae | Blues, coppers, and hair-streaks | H | N | *vegetation*; flowers, mud |
| Lyonetidae | Lyonetiid moths | H | | *leaves* |
| Noctuidae | Millers and cutworms | H | N | *vegetation*; flowers |
| Notontidae | Notodontid moths | H | N | *vegetation*; flowers |
| Nymphalidae | Brush-footed butterflies | H | N | *vegetation*; flowers, mud, dung, fruit |
| Oecophoridae | Oecophorid moths | H S | | *vegetation*; decaying material |
| Papilionidae | Swallowtails | H | N | *vegetation*; flowers |
| | | | | *vegetation*; flowers, |

| | | | | |
|---|---|---|---|---|
| Pieridae | Whites and sulfurs | H | N | mud |
| Plutellidae | Diamondback moths | H | N | *vegetation* |
| Pyralidae | Snout and grass moths | H S P | | *vegetation, decaying material*; vegetation |
| Saturniidae | Silk moths | H | -- | *vegetation* |
| Satyridae | Satyrs | H | N | *vegetation*; flowers |
| Sphingidae | Hawk and sphinx moths | H | N | *vegetation*; flowers |
| **DIPTERA** | **True flies** | | | |
| Acroceridae | Small-headed flies | Pa | N | *spiders*; flowers |
| Agromyzidae | Leaf-miner flies | H | S | *vegetation*; midair |
| Anthomyiidae | Anthomyiid flies | H P | S | *vegetation, esp. roots*; water |
| Asilidae | Robber flies | S H P | P | *soil or decaying wood*; midair, ground, flowers |
| Bibionidae | March flies | S | S | *grass roots, decaying material*; meadows |
| Bombyliidae | Bee flies | Pa P | H | flowers, streamsides |
| Calliphoridae | Blow flies | S | S | *decaying material*; dung, carrion |
| Cecidomyiidae | Gall midges | S H P | -- | *vegetation, decaying material*; midair |
| Ceratopogonidae | Sand flies | S | Pa P | *water*; midair, picnickers |
| Chamaemyiidae | Aphid flies | P | | *Homopterans*; vegetation |
| Chironomidae | Water midges | S | -- | *water*; riparian in midair |
| Chloropidae | Grass flies | H S | | *grass, decaying material*; grasses |
| Conopidae | Thick-headed flies | Pa | N | *bees, wasps*; flowers |
| Culicidae | Mosquitoes | H P | Pa H | *water*; flowers (males), victims (females) |
| Dolichopodidae | Long-legged flies | P | P | *water*, decaying material, bark; riparian vegetation |
| Drosophilidae | Fruit flies | S Pa P | H S | *decaying material*; midair, fruit |
| Empididae | Dance flies | P | P H | *soil, decaying material, water*; midair, flowers |
| Ephydridae | Shore flies | S | S | *water*; moist places |
| Hippoboscidae | Louse flies | Pa | Pa | *birds, mammals, decaying material*; moist places |
| Lauxaniidae | Lauxaniid flies | S | S | *decaying material*; moist places |
| Lonchopteridae | Spear-winged flies | S | S | *decaying material*; moist places |
| Muscidae | House flies | S | S H | *decaying material*; animals, decaying material |

| | | | | |
|---|---|---|---|---|
| Mycetophilidae | Fungus gnats | S P | H | *decaying material*; damp areas |
| Otitidae | Picture-winged flies | H S | S | *vegetation, decaying material*; vegetation |
| Phoridae | Humpbacked flies | S Pa | S | *decaying material, nests*; decaying material |
| Piophilidae | Skipper flies | S | S | decaying material |
| Pipunculidae | Big-headed flies | Pa | S | *leafhoppers*; midair |
| Psychodidae | Moth flies, sand flies | S | P | *water*; decaying material |
| Rhagionidae | Snipe flies | P | H | *soil, decaying material, water*; vegetation |
| Sarcophagidae | Flesh flies | S Pa | S | *decaying material, insect nests*; decaying material |
| Scathophagidae | Dung flies | D H | S | *dung, vegetation*; riparian veg. |
| Scatopsidae | Minute black scavenger flies | S | S | decaying material |
| Sciaridae | Root gnats | S H | S | *decaying material*; water |
| Sepsidae | Black scavenger flies | S | Pa | *decaying material*; decaying material |
| Simuliidae | Black flies | S | S | *water*; near streams, |
| Sphaeroceridae | Small dung flies | S | H | decaying material |
| Stratiomyidae | Soldier flies | H S P | H | *water*; flowers, midair |
| Syrphidae | Syrphid flies | P | Pa H | *nests, decaying material, vegetation*; flowers |
| Tabanidae | Deer and horse flies | P | N | *water*; midair, animals |
| Tachinidae | Tachinid flies | Pa | H | *insect larvae*; flowers |
| Tephritidae | Fruit flies | H | -- | *fruit*; vegetation |
| Tipulidae | Crane-flies | H | | *grass roots, decaying material,* moist places |
| Trixoscelididae | Trixoscelidid flies | | | vegetation |
| **HYMENOPTERA** | **Ants, bees, and wasps** | | | |
| Andrenidae | Andrenid bees | H | N | *burrows*; flowers |
| Anthophoridae | Carpenter bees and others | H | N | *burrows*; flowersr |
| Apidae | Bumble- and honeybees | H | N | burrows, nests; flowers |
| Argidae | -- | H | N | vegetation |
| Aulacidae | -- | Pa | N | *beetles, wasps*; flowers |
| Bethylidae | -- | Pa | N | *insect larvae*; ground, vegetation |
| Braconidae | Braconid wasps | Pa | N H | *caterpillars*; flowers, vegetation |
| Bradynobaenidae | -- | Pa | Pa | ground |
| Cephidae | Stem saw flies | H | H | *vegetation*; flowers |
| Ceraphronidae | -- | Pa | N | *insect nests;* ground, vegetation |

**APPENDIX 4: INSECTS**

| | | | | |
|---|---|---|---|---|
| Chalcididae | -- | Pa | N | *caterpillars*; flowers |
| Chrysididae | Cuckoo wasps | Pa | N | *wasp burrows*; flowers |
| Colletidae | Plasterer bees and others | H | H | *burrows, vegetation*; flowers |
| Cynipidae | Gall wasps | H | Pa | *galls*; vegetation, ground |
| Diapriidae | Diapriids | H | N | *larval flies*; moist places |
| Dryinidae | -- | Pa | N | *insects*; flowers, ground |
| Embolemidae | -- | Pa | N | *larval insects*; flowers, ground |
| Encyrtidae | -- | Pa | N | *insects*; flowers, ground |
| Eulophidae | -- | Pa | N | *insects*; flowers, ground |
| Eupelmidae | -- | Pa | N | *insects*; flowers, ground |
| Eurytomidae | Seed chalcidoids | Pa H | N | *vegetation*; flowers, ground |
| Formicidae | Ants | H C O | H P O | *nests*; ground, vegetation |
| Gasteruptiidae | -- | Pa | H | *wasp & bee nests*; flowers |
| Halictidae | Sweat bees | H | H | *nests*; flowers |
| Ichneumonidae | Ichneumonids | Pa | N | *insect larvae, spiders*; flowers |
| Leucospidae | -- | Pa | N | *insects*; flowers |
| Megachilidae | Leafcutting bees | H | H | *nests*; flowers |
| Mutillidae | Velvet ants | Pa | -- | *bee nests*; open ground |
| Mymaridae | Fairyflies | Pa | | *insect eggs*; midair |
| Ormyridae | -- | Pa | | *gall insects*; midair |
| Platygastridae | -- | Pa | N | *gall gnats*; flowers |
| Pompilidae | Spider wasps | C | H | *burrow*; flowers |
| Proctotrupidae | -- | Pa | N | *beetles, flies*; flowers |
| Pteromalidae | -- | Pa | N | *insects*; vegetation, ground |
| Sapygidae | -- | Pa | N | *bee & wasp nests*; vegetation |
| Scelionidae | -- | Pa | N | *arthropod eggs* |
| Sphecidae | Digger wasps and others | C | P N | *burrows*; ground, flowers |
| Tenthredinidae | Common sawflies | H | H P | vegetation |
| Thysanidae | -- | Pa | | *insect eggs, larvae* |
| Tiphiidae | Tiphiids | Pa | H | *insect larvae*; midair |
| Torymidae | -- | Pa H | N | *insect larvae*; midair |
| Vespidae | Paper nest wasps and hornet | C | O | *nests*; vegetation, flowers |

(1) The orders and families in this checklist have been compiled by the Bohart Museum of Entomology at UC Davis. The information on feeding style, metamorphosis, and habitat (based on Powell and Hogue 1979, Borror et al. 1989) is generalized and does not include all exceptions. The list should therefore be considered a work in progress.

(2) Feeding styles: **C** - carnivore (larva that eats animals captured by adult)

| | | |
|---|---|---|
| **H** - herbivore | **P** - predator | **Se** - seed eater |
| **N** -nectar feeder | **Pa** - parasite | **W** - wood eater |
| **O** - omnivore | **S** -scavenger | **--** - non-feeding life stage |

(3) Larval habitats are in italics, except when they are the same as for adult stages.

# Appendix 5: Fish, Amphibians, and Reptiles of Cold Canyon

| ORDER/Family | Common Name | Scientific Name |
|---|---|---|
| **SALMONIFORMES** | **TROUT, SALMON, AND RELATED SPECIES** | |
| Salmonidae | Rainbow trout | *Oncorhynchus mykiss* |
| **CAUDATA** | **SALAMANDERS** | |
| Salamandridae | California newt | *Taricha torosa* |
| Plethidontidae | Ensatina | *Ensatina escholtzi* |
| | California slender salamander | *Batrachoseps attenuatus* |
| | Arboreal salamander | *Aneides lugubris* |
| **ANURA** | **FROGS AND TOADS** | |
| Bufonidae | Western toad | *Bufo boreas* |
| Hylidae | Pacific treefrog | *Hyla regilla* |
| Ranidae | Red-legged frog | *Rana aurora* |
| | Foothill yellow-legged frog | *Rana boylei* |
| **TESTUDINES** | **TURTLES** | |
| Testudinidae | Western pond turtle | *Clemmys marmorata* |
| **SQUAMATA** | **LIZARDS AND SNAKES** | |
| Phrynosomatidae | Western fence lizard | *Sceloporus occidentalis* |
| | Sagebrush lizard | *Sceloporus graciosus* |
| | Coast horned lizard | *Phrynosoma coronatum*[1] |
| Scincidae | Western skink | *Eumeces skiltonianus* |
| Teiidae | California whiptail | *Cnemidophorus tigris* |
| Anguidae | Southern alligator lizard | *Gerrhonotus multicarinatus* |
| Boidae | Rubber boa | *Charina bottae*[1] |
| Colubridae | Ringneck snake | *Diadophis punctatus* |
| | Sharp-tailed snake | *Contia tenuis* |
| | Racer | *Coluber constrictor* |
| | Coachwhip | *Masticophis flagellum*[1] |
| | California whipsnake | *Masticophis lateralis* |
| | Gopher snake | *Pituophis melanoleucus* |
| | Common kingsnake | *Lampropeltis getulus* |
| | Long-nosed snake | *Rhinocheilus lecontei*[1] |
| | Common garter snake | *Thamnophis sertalis* |
| | Western aquatic garter snake | *Thamnophis couchi* |
| | Night snake | *Hypsiglena torquata* |
| Viperidae | Western rattlesnake | *Crotalus viridus* |

(1) Not confirmed

# Appendix 6: Birds of Cold Canyon[1]

| FAMILY<br>Common Name | Scientific Name | Seasonal Abundance[2]<br>Win | <br>Spr | <br>Sum | <br>Fall | Nesting[3] |
|---|---|---|---|---|---|---|
| **CATHARTIDAE** | | | | | | |
| Turkey Vulture | *Cathartes aura* | Uc | A | A | A | Y |
| **ACCIPITRIDAE** | | | | | | |
| Sharp-shinned Hawk | *Accipiter striatus* | Uc | R | - | Uc | - |
| Cooper's Hawk | *Accipiter cooperii* | Uc | R | R | Uc | S |
| Red-tailed Hawk | *Buteo jamaicensis* | A | A | A | A | S |
| Golden Eagle | *Aquila chrysaetos* | R | R | R | Uc | - |
| Bald Eagle | *Haliaeetus leucocephalus* | R | R | | R | - |
| **FALCONIDAE** | | | | | | |
| American Kestrel | *Falco sparverius* | C | R | R | Uc | S |
| Prairie Falcon | *Falco mexicanus* | R | R | R | R | Y |
| Peregrine Falcon | *Falco peregrinus* | Ca | Ca | Ca | Ca | S |
| **PHASIANIDAE** | | | | | | |
| California Quail | *Callipepla californica* | R | A | A | A | Y |
| Mountain Quail | *Oreortyx pictus* | R | A | Uc | Uc | S |
| Wild Turkey | *Meleagris gallopavo* | R | R | R | Uc | S |
| **COLUMBIDAE** | | | | | | |
| Rock Dove | *Columba livia* | Ca | Ca | Ca | Ca | - |
| Band-tailed Pigeon | *Columba fasciata* | Uc | R | - | - | - |
| Mourning Dove | *Zenaida macroura* | A | A | A | A | Y |
| **STRIGIDAE** | | | | | | |
| Western Screech Owl | *Otus kennicottii* | C | C | C | C | Y |
| Great Horned Owl | *Bubo virginianus* | Uc | Uc | Uc | Uc | Y |
| Long-earred Owl | *Asio otus* | R | - | - | - | - |
| Northern Pygmy Owl | *Glaucidium gnoma* | Uc | Uc | Uc | Uc | Y |
| Northern Saw-Whet Owl | *Aegolius acadicus* | R | - | - | - | - |
| **CAPRIMULGIDAE** | | | | | | |
| Common Poorwill | *Phalaenoptilus nuttallii* | - | C | C | C | S |
| **APODIDAE** | | | | | | |
| White-throated Swift | *Aeronautes saxatalis* | Uc | A | C | Uc | Y |
| Vaux's Swift | *Chaetura vauxi* | - | R | R | R | - |

| | | | | | | |
|---|---|---|---|---|---|---|
| **TROCHILIDAE** | | | | | | |
| Black-chinned Hummingbird | *Archilochus alexandri* | - | R | R | Uc | S |
| Anna's Hummingbird | *Calypte anna* | A | A | A | A | Y |
| Calliope Hummingbird | *Stella calliope* | - | R | - | R | - |
| Rufous Hummingbird | *Selasphorus rufus* | - | R | - | Uc | - |
| Allen's Hummingbird | *Selasphorus sasin* | - | R | - | - | - |
| **PICIDAE** | | | | | | |
| Acorn Woodpecker | *Melanerpes formicivorus* | Ca | Ca | - | Ca | - |
| Red-breasted Sapsucker | *Sphyrapicus ruber* | Ca | Ca | - | - | - |
| Nuttall's Woodpecker | *Picoides nuttallii* | A | A | A | A | Y |
| Downy Woodpecker | *Picoides pubescens* | R | R | - | R | S |
| Hairy Woodpecker | *Picoides villosus* | Ca | - | - | - | - |
| Northern Flicker | *Colaptes auratus* | A | Uc | R | Uc | S |
| Pileated Woodpecker | *Dryocopus pileatus* | - | Ca | - | - | - |
| **TYRANNIDAE** | | | | | | |
| Olive-sided Flycatcher | *Contopus borealis* | - | Ca | - | - | - |
| Western Wood-Pewee | *Contopus sordidulus* | - | R | - | - | - |
| Willow Flycatcher | *Empidonax traillii* | - | R | - | R | - |
| Hammond's Flycatcher | *Empidonax hammondii* | - | R | - | R | - |
| Gray Flycatcher | *Empidonax wrightii* | - | Ca | - | - | - |
| Pacific-slope Flycatcher | *Empidonax difficilis* | - | Uc | - | Uc | - |
| Black Phoebe | *Sayornis nigricans* | A | A | C | A | Y |
| Say's Phoebe | *Sayornis saya* | Ca | - | - | - | - |
| Ash-throated Flycatcher | *Myiarchus cinerascens* | - | A | C | R | Y |
| Western Kingbird | *Tyrannus verticalis* | - | Ca | Ca | - | S |
| **VIREONIDAE** | | | | | | |
| Cassin's Vireo | *Vireo cassinii* | - | R | - | Uc | - |
| Huttons's Vireo | *Vireo huttoni* | C | C | C | C | Y |
| Warbling Vireo | *Vireo gilvus* | - | Uc | Ca | Uc | S |
| **CORVIDAE** | | | | | | |
| Steller's Jay | *Cyanocitta stelleri* | Uc | Uc | - | R | S |
| Western Scrub-jay | *Aphelocoma coerulescens* | A | A | A | A | Y |
| Common Raven | *Corvus corax* | Uc | R | - | R | - |
| American Crow | *Corvus brachyrhynchos* | - | - | - | R | - |
| **HIRUNDINIDAE** | | | | | | |
| Tree Swallow | *Tachycineta bicolor* | - | Uc | C | Uc | Y |
| Violet-green Swallow | *Tachycineta thalassina* | - | R | R | - | S |
| Rough-winged Swallow | *Stelgidopteryx serripennis* | - | Ca | - | - | S |
| Cliff Swallow | *Hirundo pyrrhonota* | - | A | A | C | Y |

## APPENDIX 6: BIRDS

| | | | | | | |
|---|---|---|---|---|---|---|
| **PARIDAE** | | | | | | |
| Oak Titmouse | *Baeolophus inornatus* | A | A | A | A | Y |
| **AEGITHALIDAE** | | | | | | |
| Bushtit | *Psaltriparus minimus* | A | A | A | A | Y |
| **SITTIDAE** | | | | | | |
| White-breasted Nuthatch | *Sitta carolinensis* | R | R | - | R | S |
| **CERTHIDAE** | | | | | | |
| Brown Creeper | *Certhia americana* | - | Ca | - | - | - |
| **TROGLODYTIDAE** | | | | | | |
| Rock Wren | *Salpinctes obsoletus* | R | R | R | Uc | S |
| Canyon Wren | *Catherpes mexicanus* | Uc | Uc | C | C | S |
| Bewick's Wren | *Thryomanes bewickii* | A | A | A | A | Y |
| House Wren | *Troglodytes aedon* | R | R | R | R | S |
| **CINCLIDAE** | | | | | | |
| American Dipper | *Cinclus mexicanus* | Ca | - | - | - | - |
| **REGULIDAE** | | | | | | |
| Golden-crowned Kinglet | *Regulus satrapa* | R | - | - | R | - |
| Ruby-crowned Kinglet | *Regulus calendula* | A | C | - | C | - |
| **SYLVIIDAE** | | | | | | |
| Blue-gray Gnatcatcher | *Polioptila caerulea* | - | Uc | Uc | Uc | Y |
| **TURDIDAE** | | | | | | |
| Western Bluebird | *Sialia mexicana* | Ca | Ca | - | Ca | S |
| Townsend's Solitaire | *Myadestes townsendi* | Ca | - | - | - | - |
| Swainson's Thrush | *Catharus ustulatus* | - | R | - | - | - |
| Hermit Thrush | *Catharus guttalus* | A | Uc | - | A | - |
| American Robin | *Turdus migratorius* | A | C | - | Uc | - |
| Varied Thrush | *Ixoreus naevius* | R | - | - | R | - |
| **TIMALIIDAE** | | | | | | |
| Wrentit | *Chamaea fasciata* | A | A | A | A | Y |
| **MIMIDAE** | | | | | | |
| Northern Mockingbird | *Mimus polyglottus* | - | Ca | Ca | Ca | S |
| California Thrasher | *Toxostoma redivium* | C | C | C | C | Y |
| **STURNIDAE** | | | | | | |
| European Starling | *Sturnus vulgaris* | R | R | - | R | S |

| | | | | | | |
|---|---|---|---|---|---|---|
| **BOMBYCILLIDAE** | | | | | | |
| Cedar Waxwing | *Bombycilla cedrorum* | Ca | - | - | Ca | - |
| **PTILOGONATIDAE** | | | | | | |
| Phainopepla | *Phainopepla nitens* | - | Ca | - | - | S |
| **PARULIDAE** | | | | | | |
| Orange-crowned Warbler | *Vermivora celata* | Ca | A | - | C | S |
| Nashville Warbler | *Vermivora ruficapilla* | - | Uc | - | Uc | - |
| Yellow Warbler | *Dendroica petechia* | - | Uc | - | Uc | - |
| Yellow-rumped Warbler | *Dendroica coronata* | A | C | - | C | - |
| Black-throated Gray Warbler | *Dendroica nigrescens* | - | C | Ca | Uc | S |
| Townsend's Warbler | *Dendroica townsendi* | - | R | - | R | - |
| Hermit Warbler | *Dendroica occidentalis* | - | R | - | Uc | - |
| MacGillivray's Warbler | *Oporornis tolmiei* | - | R | - | - | - |
| Wilson's Warbler | *Wilsonia pusilla* | - | Uc | Ca | Uc | - |
| **THRAUPIDAE** | | | | | | |
| Western Tanager | *Piranga ludoviciana* | - | Uc | - | Uc | - |
| **EMBERIZIDAE** | | | | | | |
| Spotted Towhee | *Pipilo maculatus* | A | A | A | A | Y |
| California Towhee | *Pipilo crissalis* | A | A | A | A | Y |
| Rufous-crowned Sparrow | *Aimophila ruficeps* | Uc | Uc | Uc | Uc | S |
| Sage Sparow | *Amphispiza belli* | Uc | Uc | Uc | Uc | S |
| Chipping Sparrow | *Spizella passerina* | - | Ca | - | - | - |
| Black-chinned Sparrow | *Spizella atrogularis* | - | Ca | - | - | - |
| Lark Sparrow | *Chondestes grammacus* | - | Ca | - | - | - |
| Fox Sparrow | *Passerculus iliaca* | C | Uc | - | Uc | - |
| Song Sparrow | *Melospiza melodia* | R | - | - | R | - |
| Lincoln's Sparrow | *Melospiza lincolnii* | Ca | - | - | - | - |
| White-throated Sparrow | *Zonotrichia albicollis* | Ca | - | - | - | - |
| Golden-crowned Sparrow | *Zonotrichia altlricapilla* | A | C | - | - | - |
| White-crowned Sparrow | *Zonotricha leucophrys* | A | C | - | C | - |
| Dark-eyed Junco | *Junco hyemalis* | A | Uc | - | C | - |
| **CARDINALIDAE** | | | | | | |
| Black-headed Grosbeak | *Pheucticus melanocephalus* | - | C | C | R | - |
| Lazuli Bunting | *Passerina amoena* | - | Ca | Ca | - | Y |

| ICTERIDAE | | | | | | |
|---|---|---|---|---|---|---|
| Western Meadowlark | *Sturnella neglecta* | Ca | - | - | Ca | - |
| Brewer's Blackbird | *Euphagus cyanocephalus* | Uc | Uc | Uc | Uc | Y |
| Brown-headed Cowbird | *Molothrus ater* | - | Ca | - | - | S |
| Bullock's Oriole | *Icterus bullocki* | - | Uc | Uc | R | Y |
| **FRINGILLIDAE** | | | | | | |
| Purple Finch | *Carpodacus purpureus* | Uc | - | - | Uc | - |
| House Finch | *Carpodacus mexicanus* | A | A | A | A | Y |
| Pine Siskin | *Carduelis pinus* | Ca | - | - | - | - |
| Lesser Goldfinch | *Carduelis psaltria* | A | A | A | A | Y |
| Lawrence's Goldfinch | *Carduelis lawrencei* | - | Ca | - | - | - |
| American Goldfinch | *Carduelis tristis* | C | Uc | - | Uc | S |

(1) The checklist reflects observations made by many people between 1975 and 2000, with fairly extensive surveys by Andrew Engilis, Jr., Cameron Barrows, Katherine Barrows, Ron Cole, Sid England, and Wesley Weathers. In addition, a number of casual birders and also more organized outings (e.g., the Audubon's Christmas bird count) have contributed to this table. Starting in 1976, the Cold Canyon basin has been included within the Audubon Christmas bird count circle (Area 10). Records from these counts are published in Burrowing Owl, the official publication of the Davis Audubon Society.

(2) Because species abundance is so seasonal, the checklist above includes estimates for seasonal abundance, abbreviated as follows:

| Symbol | Category | Observability | Meaning |
|---|---|---|---|
| **A** | Abundant | 100-76% | Hard to miss |
| **C** | Common | 75-51% | Should see |
| **Uc** | Uncommon | 50-26% | May see |
| **R** | Rare | 25-11% | Hard to see |
| **Ca** | Casual | 10-1% | You'll be very lucky |

This classification ignores "fly overs," e.g., waterfowl and gulls that do not actually land in the canyon.

(3) **Y** - confirmed nesting, **S** - suspected nesting.

# Appendix 7: Mammals of Cold Canyon

| ORDER/Family | Common Name | Scientific Name |
|---|---|---|
| MARSUPIALIA | MARSUPIALS | |
| Didelphi | Virginia opossum | *Didelphis virginiana* |
| | | |
| INSECTIVORA | INSECTIVORES | |
| Talpidae | Broad-handed mole | *Scapanus latimanus* |
| Soricidae | Trowbridge shrew | *Sorex trowbridgii* |
| | Vagrant shrew | *Sorex vagrans* |
| | Ornate shrew | *Sorex ornatus* |
| | | |
| CHIROPTERA | BATS | |
| Vespertilionidae | Yuma myotis | *Myotis yumanensis* |
| | Long-eared myotis | *Myotis evotis* |
| | Long-legged myotis | *Myotis volans* |
| | California myotis | *Myotis californicus* |
| | Western pipistrelle | *Pipistrellus hesperus* |
| | Big brown bat | *Eptesicus fuscus* |
| | Western red bat | *Lasiurus blossevilii* |
| | Hoary bat | *Lasiurus cinereus* |
| | Townsend's long-eared bat | *Corynorhinus townsendii* |
| | Pallid bat | *Antrozous pallidus* |
| Molossidae | Brazilian free-tailed bat | *Tadarida brasiliensis* |
| | | |
| CARNIVORA | CARNIVORES | |
| Ursidae | American black bear | *Ursus americanus* |
| Procyonidae | Ringtail | *Bassariscus astutus* |
| | Raccoon | *Procyon lotor* |
| Mustelidae | Long-tailed weasel | *Mustela frenata* |
| | American mink | *Mustela vison* |
| | American badger | *Taxidea taxus* |
| Mephitidae | Striped skunk | *Mephitis mephitis* |
| | Western spotted skunk | *Spilogale gracilis* |
| Canidae | Coyote | *Canis latrans* |
| | Gray fox | *Urocyon cinereoargenteus* |
| | Red fox | *Vulpes vulpes*[1] |
| | Domestic (feral) dog | *Canis familiaris*[1] |
| Felidae | House (feral) cat | *Felis silvestris* |
| | Bobcat | *Lynx rufus* |
| | Mountain lion | *Puma concolor* |
| | | |
| RODENTIA | RODENTS | |
| Sciuridae | California ground squirrel | *Spermophilus beecheyi* |
| | Sonoma chipmunk | *Tamias sonomae* |

**APPENDIX 7: MAMMALS**

| | | |
|---|---|---|
| | Western gray squirrel | *Sciurus griseus* |
| Geomyidae | Botta's pocket gopher | *Thomomys bottae* |
| Heteromyidae | California pocket mouse | *Perognathus californicus*[1] |
| | San Joaquin pocket mouse | *Perognathus inornatus* |
| | California kangaroo rat | *Dipodomys californicus* |
| Muridae (Criticinae) | Deer mouse | *Peromyscus maniculatus* |
| | Brush mouse | *Peromyscus boylii* |
| | Pinyon mouse | *Peromyscus truei* |
| | Harvest mouse | *Reithrodontomys megalotis* |
| | Dusky-footed wood rat | *Neotoma fuscipes* |
| Muridae (Arvicolinae) | California vole | *Microtus californicus* |
| Muridae (Murinae) | House mouse | *Mus musculus* |
| LAGOMORPHA | RABBITS | |
| Leporidae | Black-tailed hare | *Lepus californicus* |
| | Audubon's cottontail | *Sylvilagus audubonii* |
| | Brush rabbit | *Sylvilagus bachmani* |
| ARTIODACTYLA | EVEN-TOED HOOFED MAMMALS | |
| Suidae | Feral hog | *Sus scrofa* |
| Cervidae | Mule deer | *Odocoileus hemionus* |

(1) Not confirmed

# References


## General

The American River Natural History Association. 1993. *The Outdoor World of the Sacramento Region, a Local Field Guide*. The American River Natural History Association, Sacramento, CA.

Weathers, W.W., and R. Cole, eds. 1985. *Flora and Fauna of the Stebbins Cold Canyon Reserve, Solano County, California*. Institute of Ecology, University of California, Davis, CA.

## History

Barrett, S.A. 1908. The ethnography of the Pomo and neighboring Indians. *University of California Publications in American Archeology and Ethnology*, 6:1-332.

Begell, R., and M. Guida. 1995. The Solano Project and Monticello Dam. *The Solano Historian*. Vol. XI, no. 1. pp. 6-7.

Kroeber, A.L. 1930. The Patwin and their neighbors. *University of California Publications in American Archeology and Ethnology*, 29:253-428.

Lange, D., and P. Jones. 1960. Death of a valley. *Aperture*, Vol. 8, no. 3.

Larkey, J.L. 1991. *Winters: a Heritage of Horticulture, a Harmony of Purpose*. Yolo County Historical Society, Woodland, CA.

Rubin, H. 1988. *The Solano Water Story: A History of the Solano Irrigation District and the Solano Project*. Solano Irrigation District, Vacaville, CA.

## The Physical Environment

Alt, D.D., and D.W. Hyndman. 1975. *Roadside Geology of Northern California*. Mountain Press, Missoula, MT.

McPhee, J.A. 1993. *Assembling California*. Farrar, Straus & Giroux, New York.

Oqvist, C. 1999. *Surface and Bedrock Geology of the Cold Canyon Area, Solano County, California*. Undergraduate Thesis, University of California, Davis, CA.

## Plants

Elias, T.S., and P.A. Dykeman. 1982. *Edible Wild Plants: A North American Guide*. Outdoor Life Books, New York.

## REFERENCES

Hickman, J.C. 1993. *The Jepson Manual: Higher Plants of California*. University of California Press: Berkeley.

### Invertebrates

Borror, D.J., C. A. Triplehorn, and N. F. Johnson. 1989. *An Introduction to the Study of Insects*. 6th ed. Saunders College Publishing, Philadelphia.

Chinery, M. 1993. *Spiders*. Whittet Books, London.

Kaston, B.J., and E. Kaston. 1953. *How to know the spiders*. W.C. Brown Publishers, Dubuque, IA.

Powell, J.A., and C.L. Hogue. 1979. *California Insects*. University of California Press, Berkeley.

Roberts, M.J. 1995. *Spiders of Britain and Northern Europe*. HarperCollins, London.

### Vertebrates

Burt, W.H. and R.P. Grossenheider. 1952. *A Field Guide to the Mammals*. Houghton Mifflin, Boston.

Ehrlich, P.R., D.S. Dobkin, and D. Wheye. 1988. *The Birder's Handbook*. Simon and Schuster, New York.

Ingles, L.G. 1965. *Mammals of California*. 3rd ed. Stanford University Press, Palo Alto, CA.

National Geographic Society. 1987. *Field Guide to the Birds of North America*, 2nd ed. National Geographic Society, Washington, DC.

Peterson, R.T. 1990. *A Field Guide to Western Birds*, 3rd ed. Houghton Mifflin, Boston.

Stebbins, R.C. 1972. *California Amphibians and Reptiles*. University of California Press, Berkeley.

Stebbins, R.C. 1985. *A Field Guide to Western Amphibians and Reptiles*. 2nd ed. Houghton Mifflin, Boston.

Storer and Usinger. 1963. *Natural History of the Sierra Nevada*. University of California Press, Berkeley.

# Index

This index covers names, places, and technical terms covered in the chapters, but not in the appendices or references. Boldface numbers indicate illustrations or photos.

## INDEX

## INDEX

## INDEX

# Acknowledgements

The production of this manual could not have proceeded without the help of a number of people. Susan Harrison, Sandy Harcourt, and Virginia Boucher secured the funding that made this work possible. Virginia Boucher, Yvonne Cardenas, Jennifer French, Sandy Harcourt, Susan Harrison, Doug Kelt, Eldridge Moores, Dirk Van Vuren, and Phil Ward provided insightful comments on early drafts. Lynn Kimsey, Art Shapiro, John DeBenedictis, and Jim Fordyce greatly improved the invertebrate chapter and appendices. Asfala Arison provided invaluable assistance with layout. Dan Tolson provided logistical support and much-needed enthusiasm. Tim Herrlinger permitted use of computers to produce the layout. Victoria Gonzales, Meghan Kelly, and Ed SantaAna at DeHart's were very helpful with the printing.

Mike Benard, Virginia Boucher, Ellen Dean, Sandy Harcourt, Kate Mawdsley, Theresa Talley, and Shane Waddell provided edits for the second edition.

We also thank the following people, groups, and publishers for permission to reprint illustrations and photographs:

McGraw-Hill Companies, Inc. – six drawings of spiders and their webs from Kaston and Kaston (1953)

Peter Moyle – rainbow trout (p. 67)

Courtney Oqvist – mineral spring (p. 15) and geologic map (p. 94)

Marcel Rejmanek – G. Ledyard Stebbins (p. 12)

Michael J. Roberts and HarperCollins Publishers – ten drawings of spiders and their webs from Roberts (1995)

Jo Smith and Camilla Gunderson, American River Natural History Association – numerous drawings of animals from *The Outdoor World of the Sacramento Region*

Solano County Historical Society – dam construction (p. 11)

The Solano Irrigation District – Devil's Gate (p. 11)

University of California Press – numerous drawings from Storer and Usinger (1963) and from Powell and Hogue (1979)

Whittet Books – scorpion (p. 47) and harvestman (p. 49) from Chinery (1993).